THE REAL

APP

THE REAL ASTROLOGY APPLIED

Teachings from the Astrologer's Apprentice

John Frawley

APPRENTICE BOOKS

Published 2001 by Apprentice Books

www.johnfrawley.com

ISBN 978-0-9539774-1-3

Cover photograph by Sergio Bondioni, Yellow Brick Studios
Design and typesetting by John Saunders Design & Production

Contents

Introduction

Turn left at the Twenty-first Century, follow the rutted pathway through the woods, and you will soon arrive at the village on whose outskirts stands our workshop.

Passing by the cluster of would-be apprentices, kept waiting for a day or two outside the gate in order to weed out those with only little desire to learn, you will meet first old Amos, our gatekeeper. Given half a chance – or even no chance at all – he will regale you with stories of his youth, when he, our Master and William Lilly set out together on the highway leading to knowledge of the celestial art. An expression of urgent determination will usually serve to release you from his reminiscences in a matter of hours.

The building immediately before you as you enter our yard is the stables, where the Sagittarius and other bestial signs are tended by the stable-lads – an unruly bunch, but useful. (The Aquarius, Virgo and other humane signs will soon be housed in en-suite accommodation in the new block that you may see rising just beyond the stables; but for the time being they must share a blanket and some floor space with the less insalubrious of the apprentices.) On your right is the forge, where Robin, our smith, with imperious stroke batters out the house-cusps that will be bolted together to make charts. You will often see Ned and Colin, our carpenters, silhouetted against the flames there, as it is their habit, especially in winter, to sit for warmth and company in the forge as they carve the planetary glyphs that will be set into these charts.

In the yard itself you might see some planets stretching their legs while their chart is rectified, or perhaps hear the raucous din from the antiscia enclosure as feeding time draws near. Apprentices scurry about as they have been told, and in quiet times you may even see our Master himself enjoying a pipe beneath the mulberry tree.

To your left is the scriptorium, where at our Master's behest for the past few years I and a group of scribes have produced the magazine *The*

Astrologer's Apprentice, from which most of the articles in this collection are drawn. But as an honoured guest you will be hurried past these sights into the refectory, where you will be greeted, as is our custom, with a jug of old ale and a slab of cheese. Welcome to our world!

The Apprentice

Preface

This is a collection of articles that deal in greater depth with subjects raised in *The Real Astrology.*

Most of these articles appeared first in the journal *The Astrologer's Apprentice.* As its masthead proclaims, *The Apprentice* is 'The Tradition as it Lives', with detailed articles on both theory and practice demonstrating that Real Astrology is alive and well – if somewhat neglected - even in the Twenty-first Century.

Two articles, *Mutual Reception: Our Magic Wand* and *The Internal King,* were published in *The Astrological Association Journal.* The series on the houses is from *Horoscope* (UK edn.). While this series makes no claim to being the definitive treatment of the subject, it has occasioned many requests for reprints and so is included here 'by public demand'. All articles have been slightly revised for this edition.

I trust that my gentle reader will forgive a certain repetition that is inevitable in a collection such as this: various paths up the same mountain will give similar views from time to time. Approaching the same point from different directions can only add to understanding.

Acknowledgements

Victor Laude has, as ever, given freely of his time and wisdom. Whatever is of truth in these pages owes much to him; what is of error, nothing.

Despina Giannakopoulou has played an invaluable part in the genesis of this book. Margaret Cahill and Branka Stamenkovic have made helpful contributions at the editing stage. The persistence of my students in demanding elucidation of what the texts leave obscure has been the inspiration for many of the articles republished here. It would be invidious to single out individuals: to them all, my thanks.

My family has been as patient as ever – Olivia insists upon a special mention. Seth, Amos, Diggory, and even the stable-lads have rendered assistance after their own fashion. Special thanks are due to the inimitable Neptunia, whose contribution has, of course, been unique, and finally to those who have supported *The Astrologer's Apprentice* since its birth in 1996.

Key

♈	Aries	ruled by Mars
♉	Taurus	ruled by Venus
♊	Gemini	ruled by Mercury
♋	Cancer	ruled by the Moon
♌	Leo	ruled by the Sun
♍	Virgo	ruled by Mercury
♎	Libra	ruled by Venus
♏	Scorpio	ruled by Mars
♐	Sagittarius	ruled by Jupiter
♑	Capricorn	ruled by Saturn
♒	Aquarius	ruled by Saturn
♓	Pisces	ruled by Jupiter

♄	Saturn
♃	Jupiter
♂	Mars
☉	Sun
♀	Venus
☿	Mercury
☽	Moon

☊	North Node of the Moon
☋	South Node of the Moon
⊗	Part of Fortune/Fortuna

☌	Conjunction	same degree, same sign
☍	Opposition	same degree, opposite sign
△	Trine - 120 degrees	same degree, 4th sign round
□	Square - 90 degrees	same degree, 3rd sign round
⚹	Sextile - 60 degrees	same degree, 2nd sign round
℞	Retrograde	appears to be going backwards

Conventions

For simplicity, the terms 'Lord 1, Lord 2, etc.' are used here to mean 'the planet that rules the sign on the cusp of the first house, second house, etc.'

Horary charts are cast using Regiomontanus cusps; all other charts using Placidus.

Footnotes added for this edition are in square brackets.

Following the example of the astrological authorities on which we rely, it is the policy of *The Astrologer's Apprentice* not to cite references unless there is particular reason for so doing. This policy is continued in this volume. Were references to be cited for our more technical articles (not included here) the text would drown in a tide of footnotes. More importantly, I do not feel that astrology is flattered by being dressed in the robes of academia. My hope here is that readers may be inspired, not to pick cherries, but to buy the orchard for themselves.

1

In At The Deep End

ARE YOU RECEIVING ME?

The merciless midday Sun beat down on that square of parched earth in some far-flung corner of empire, glinting off the teeth of the grinning infidel taking guard in front of his wicket. “Help me off with this dashed plaster-cast, Tomkinson; it impedes my run-up,” the gallant bowler instructed his valet.

“But your triple compound fracture, sir,” he queried.

“Can’t think about that now, man; the honour of the regiment is at stake.”

So saying, Walmsley ‘Johnny’ Fitzwarren (Lieut. the Rt. Hon., D.F.C. and bar), a trickle of sweat snaking down his forehead, tossed back his flaxen forelock and took hold of the ball with that special grip invented all those years back over muffins after a hard afternoon’s fagging. “Chin up,” he whispered to himself, preparing for that patent under-googlied off-break that not even ‘Nobby’ Royston-Davies, Head of School and opening bat for the first XI, had ever quite been able to master.

With barely a flicker of his under-lip betraying the pain of running on two broken legs, he loped up to the popping-crease, sending the ball spinning on its elusive way. But, easing his Luger automatic in its holster and pausing only to kick a passing dog, the batsman lofted it effortlessly away over covers for six more runs.

As he struggled to fit his artificial arm back into its socket, from whence it had slipped as he bowled, ‘Johnny’ felt his captain’s hand on his shoulder. “You’re the last man left alive on our team, Johnny – it’s up to you. Line and length, old boy; line and length.”

And thus has it always been: line and length, line and length. Forget all the fancy stuff; just get the ball going in the right direction, for the right distance, and all things will come to you.

And, as cricket is ever one of the profounder metaphors for life, thus it is in astrology as well: line and length, line and length. When light is fading on a sticky wicket, and some particularly deranged client pops up with an even more unbelievable question, rather than panicking or wondering if asteroids have anything to do with it, break it down to basics and all will be revealed before you.

These basics are threefold: aspect, dignity and reception. All that you need is there. If the consideration of these three points doesn't yield an answer, the solution is simple – go back and consider them some more. You may need to bowl an awful lot of balls, but if you keep getting the line right and the length right, you will eventually get him out.

Listen carefully, I will say this only once...

Each astrological chart is like a drama, with the planets a stock company of players. When we have identified which of the planets represents each of the characters in our drama, we then look to see what they are up to:

Dignity shows their power to act.
Reception shows their inclination to act.
Aspect shows the occasion to act.

The text-books concentrate heavily on aspect. They go into considerable detail on dignity. They talk little of reception. Lilly, for example, in his theoretical pages, gives it scant mention, a scantiness that has led those whose careful study of Lilly extends not quite so far as reading his work, to conclude that he makes but little use of it. But turn to his worked examples; follow his logic; pull them apart, asking continually "Why is he doing this?" and it becomes clear that the careful consideration of reception was fundamental to his method. Even though he fails to shout it from the roof-tops, chart after chart makes no sense without it.

This failure to explain its use in detail is probably because, to the astrologer who is living within the tradition rather than attempting to comprehend it from outside, reception is just common sense. This is true of all astrology and underlines the importance of at least an understanding of the tradition, if not – consummation devoutly to be wished – standing within it. For the scientists are quite correct: viewed from within the modern mindset, astrology is utter nonsense, and all the distortions through which modern astrologers put our art in attempt to convince themselves otherwise alter this fact not one jot: astrology is rubbish. Viewed from within the traditional mindset, astrology is nothing but the application of common sense. Accept the premises and astrology follows naturally and inevitably – which brings us back to 'line and length'.

Reception is to such an extent just 'common sense' that most writers on astrology never found it worth taking time to describe it. But the world has changed, so here it is.

According to the tradition "love is all there is, it makes the world go round; love and only love, it can't be deni-i-i-ied"; and when we find Aristotle and Bob Dylan in agreement, we must surely take notice. This concept of love, however, is something far broader and grander than our immediate thought of romantic affection; this is love in the sense that its constant outpouring of energy shows that the Sun loves the Earth. The tradition, in whichever of its branches we find it, is quite clear on this matter. "God is love," we are told; the insertion of a copulative that is not present in the semitic languages diluting the power of the image, making it seem like "God is quite fond of us," rather than the direct and literally true "God:Love" of the original.

Everything's motive for doing anything, we are told, is love, and the receptions in the chart show us exactly what it is that all our various characters do love, and in what order of priority they place them. Each planet falls in a selection of different dignities – sign, triplicity, terms, face, and often exaltation. Each of these dignities is ruled by one or other of the planets. The planet loves whichever planets rule the dignities in which it falls. And it loves them in different ways, depending upon the nature of that particular dignity.

Thinking of reception in terms of love is simple when dealing with relationship questions: "Yes, Jimmy loves Jane; yes, she fancies the hunk in accounts." It can also be applied to all other questions. If I am asking, for instance, "When will I get paid?" it is reassuring to find that my boss's money loves me. If it loves me, it wants to be near me, so this is a step in the right direction. If my boss's money loves only my boss, however, and especially if this exclusive passion is reciprocated, I am unlikely to win its hand. Similarly, I hope to see that the car I am thinking of buying, my new house and the medicine I have been prescribed are all suitably fond of me. If, for example, the medicine were in my significator's detriment, I would be concerned that it meant me no good.

A planet loves truly that planet which rules the sign in which it falls. It sees it, with all its faults; understands it; knows it thoroughly. This is its prime true interest, although this can be overshadowed by others. So if the significator of the quesited is in my sign, it loves me – whatever it is: person, house, job, money.

A planet exalts the planet in whose exaltation it falls. This is powerful, but is never quite real. It does not see it clearly. Charts cast in the early stages of relationships usually show receptions by exaltation: the aura of divinity has not yet been pierced by knowledge of his unsavoury personal habits. Charts for "Is it really over?" questions frequently show significators which have recently moved out of reception by exaltation.

Given support by some other dignity – triplicity is a good one – reception by exaltation can sustain a relationship. It will not be one in which the partners slob around on the sofa, scratching themselves and eating cold pizza; it will be necessary to work at sustaining the illusion (an ability to put on make-up before awaking would be a useful gift), but it is viable. Similarly with other things than relationships: if my job exalts me (Lord 10 in exaltation of Lord 1), it likes me, but over-estimates my abilities; if I exalt it (Lord 1 in exaltation of Lord 10), there is a warning that it is unlikely to meet my unrealistic expectations.

Triplicity is the reception of friendship. This is why it supports exaltation so well: exaltation plus friendship will make a workable substitute for true love. A planet in its own triplicity is, quite literally, 'in its element': it is comfortable, well able to cope. If it is in another planet's triplicity, it is comfortable, at ease with, whoever that planet signifies. So if Jimmy and Jane's planets are in mutual reception by triplicity, there is no grand passion, but they are the best of friends. If I and my job are in mutual reception by triplicity, it is not my ideal vocation, and I will not become one of the heroes of that trade's mythology, but I am quite content doing it and it is well within my capacity.

Terms and face are much more minor; they are also more fleeting than the other dignities, as they cover only part of a sign. Allied with one of the other dignities, they can be powerful; by themselves, they show only small interest – but this is a good deal better than none at all. In context, this may be quite acceptable: if I were applying for a minor job with a big company, finding the job's planet in my terms or face would be a pleasant surprise – I am lucky it cares about me even that much. If I were applying to be Managing Director, I would feel slighted: it clearly has someone else in its life.

So, while the planet's own dignity or debility will show whether it has power to act, and the aspects it forms will show if the occasion to act will arise, the receptions show what its priorities are, and therefore if it wants to act. It matters little if Susie sits beside me on the bus, even if my charms are at their most potent, if she is besotted with someone else; this would be clearly shown in the chart.

As a planet's receptions show what it loves, they show also what has power over it: this can sometimes be a more useful way of looking at them. There is no stronger sign of love in the chart than finding the querent's planet just inside the seventh house, the house of the beloved. The planet will then, of course, be in the sign of the quesited – showing love – and also in its own detriment, indicating its own weakness. This is partly love vaunting not itself; but it also reflects the amount of power that the beloved has over the lover. In other placements, the degree of power is shown by the kind of reception.

When judging the chart, we will usually have an idea of the degree of reception we hope to find, depending on the question. If it were "Should I go ahead and marry him?" we would hope to see mutual reception, and something rather stronger than mutual reception by face. If the question were "Will our business partnership succeed?" however, we might find no reception at all; but if both planets were disposited by the same third planet, and that third planet signified something relevant – the business's profit, perhaps – we might well judge that to be a sound basis for a business. OK, the partners don't like each other much, but they both want to make a profit for the company. While we will have an idea of the reception we want to find, we must also make the effort to understand whatever reception is actually there: we cannot dismiss it because it doesn't fit our preconceptions.

Mixed receptions – that is, mutual receptions involving different dignities – are perfectly acceptable, and describe the situation. Suppose Mars is the boy and Venus the girl. Mars in Libra shows that he loves her, fairly clear-sightedly. Venus might be at 20 Capricorn: she exalts Mars, and is in his terms. So she is very keen on him, but has him on a pedestal; being in his terms shows that, beside the adulation, there is also a degree of real liking. Especially if Saturn, the ruler of Capricorn, were in strong dignities of Venus, we might want to find out what Saturn represents, as it is clearly of great importance to her. In this instance, with Mars in Saturn's exaltation and, if it is a day chart, his triplicity as well, it is also of importance to Mars. Maybe it is a third party, perhaps an ex-partner who still retains great influence on the situation; maybe it is their relationship (Saturn might be dispositor of the Part of Marriage, for instance); or maybe it is something else: the chart will tell us.

The close and careful study of what receptions each significator has, and what the planets that receive it signify, will enable us to unravel the knottiest of situations and even – in far more concrete terms than those jung astrologers still greene in judgement – the most subtle psychological problems. And the more we understand these receptions, the more perforce we will become aligned with the tradition behind our art; and the more we do that, the better astrologers we shall become.

C'est tellement simple, l'amour.

Peut-être pour toi, Garance. But not, to our bank-manager's relief, pour tout le monde. So we shall now consider briefly exactly what means what when judging relationship horaries. We have seen how receptions will reveal each person's true attitude; to complete the picture, we need to understand the specific meanings of the various significators.

The Lord of the Ascendant shows the querent; the Lord of the seventh, the quesited. The fifth house has nothing to do with it. The fifth shows the things people do during relationships: eating out, running hand-in-hand through the waves, having sex; it does not show the person involved, no matter how casual a relationship it might be.

The planets ruling the first and seventh houses show the people as individuals. The Moon always co-rules the querent, unless it is main significator of the quesited, in which case they have first claim on its services. The Moon shows the querent's feelings. So a comparison of the dignities and relationships of the Ascendant ruler with those of the Moon will tell us a great deal about the querent's attitude, particularly whether heart and mind are in agreement.

Venus co-rules the woman, the Sun co-rules the man. Not Mars: Mars usually signifies sex, unless there is a specific context in the chart which gives it a prior significance (such as being main significator of one of the parties involved). Not surprisingly, there are usually plenty of Mars receptions in charts cast during the early stages of relationships: often, there is little reception between the main significators, but both will share powerful Mars dignities. It is also a popular choice in "My husband has a bit on the side; should I murder him now?" questions: if husband and floozy are in mutual reception, wife has cause for concern; if they just share Mars dignities, he will soon be back.

If they are being used as significators of Man or Woman, the Sun and Venus signify the people very much as animal beings: Me Tarzan, You

Jane. So if the receptions are heavily biased towards Sun/Venus, we see strong physical attraction, or the biological imperative to associate with some reasonably tolerable member of the opposite sex. Sun and Venus are not notably fussy. This does not apply if they are in use as rulers of the first or seventh houses.

We also have secondary significators, such as the planet from which the Moon is separating (querent) and that to which it applies (quesited), for emergency use only. Saturn, if it is not doing anything else, can often show an ex-partner, or someone the querent devoutly wishes were an ex-partner. In the ideal chart, of course, all the various significators unite harmoniously: body, heart and head all pull in the same direction. In practice, if this is the case our potential querent is far too busy billing and cooing to ask horary questions. It is by considering which planets are engaged and how they relate by aspect and reception that we can judge exactly what is happening, and therefore what is to come.

MUTUAL RECEPTIONS: OUR MAGIC WAND

The common image of a consultation with an astrologer is like a scene from a horror movie. In semi-darkness and to the accompaniment of some suitably eerie music the client nervously enters the hallowed ground of the astrologer's den, preferably accompanied by a goat in case a sacrifice is required. There, a gaunt figure with staring eyes reveals intricate and unknown details about past and present, to the sound of gasps of amazement from the client, who sits tight-lipped, revealing not one word about his situation in life.

As most readers will know, this image is far from the truth. Even the bat-haunted lair of the traditional astrologer is not quite like this. Disappointing though it may be, reality is far more mundane, and any steaming goblet offered our client is more likely to contain a tea-bag than eye of newt or tooth of dog.

There are those clients who sit in obstinate silence, expecting to be astounded. We can ask them if they go to their doctor, demand "Tell me about myself," and then sit mute as he scrapes and pokes around their body, looking for clues to their present situation, past ailments and diseases that they might contract in the future. With practice and experience the doctor could tell a good deal about them even if they were this uncooperative – if, for example, they had been brought in unconscious. But it simplifies matters no end if they talk to him.

Similarly with an astrologer. "Can you read my birth-chart?" is equivalent to asking: "Can you tell me about my life?" Yes, the astrologer can tell you about your life; but it does help if he has some clue as to which facet of your life you would like to examine. If you have a stomach-ache, there is little point in the doctor looking into your ears. Similarly, if you want to know if Cedric or Algernon is the man for you, the astrologer is wasting his time working out how many brothers and sisters you might have. All the information about siblings, parents, friends and childhood is there in the chart, waiting to be revealed; but while telling these things can be an impressive trick, there is little point in the astrologer working for hours to tell you things that you could tell him in seconds.

So what is the purpose of the reading? There is undeniably a certain value in being confronted with things in black-and-white. We all have our little foibles that we are convinced are utterly endearing, but which the birth-chart unequivocally shows to be destructive to ourselves or to our relationships with others. This is all the more true as television and films increasingly reinforce a strange view of life. It can be salutary to be informed, for example, that your drinking does not make you a romantic hell-raiser, just a drunken slob.

There is, then, a purpose in using the reading to understand yourself and what is happening in your life, either overall or, by studying progressions and transits, at that particular time. The obvious example of this is in dealing with the fearsome Saturn return. How often do we have clients telephone saying: "I don't know what's going on in my life. I've lost all direction. I feel so unsure of myself and generally miserable – not like myself at all."

"So you're about 28-29 years old?" the all-wise astrologer replies.

"How did you know that!" comes the astonished response.

Because, of course, it is at that age that everyone experiences their first Saturn return. It can be immensely reassuring to understand that this is a natural stage in the life, that it happens to everyone with varying degrees of intensity and duration, and that it will soon pass.

Understanding what is going on, however, is a largely passive experience. If I am about to be run over by a steam-roller, I am less concerned with understanding the exact mechanics of what will happen to me, than I am in knowing what I might be able to do to avoid it happening. Often, then, the purpose of the astrological consultation is to arrive at some positive intervention that can be made in the life. But how do we do this?

One possibility is by the construction of talismans. As with all forms of astrological magic, however, the great danger with this is that it might actually work. The lesson which the practice of horary makes distressingly clear – and this is one reason for horary being rightly considered the gateway into astrology – is 'Be thankful we don't get what we think we want!' If granted our desires, we are far too likely to end up like the hero in the fairy-tale, begging the genie to put things back as they were before.

Astrological influence cannot be turned on and off like a tap.

Consider: our young bravo has just received his call-up papers, and worries that his courage might fail him in the heat of battle. So we perk up his Mars by giving him a talisman. Thus perked up, his Mars involves him in a brawl at barracks, and he is dead before he ever reaches the front line. Extreme caution and subtlety of knowledge is required.

Much can be done with the less drastic tools of diet and precious stones. Imaging, role-play, alterations in dress, taking up or dropping certain activities: these are all of great use, and answer the astrologer's frequent puzzled question: "OK, I've identified the issue; but what do I do now?" But having all these resources is one thing; we also need to know how and when to apply them. The problem is that the planet – or planets – that is causing the difficulty can often not be tackled head-on without exacerbating the problem ("You need to stop arguing." "No I don't!"). More productive is to sneak up on it from behind. To do this, we need to study the receptions in the chart and, most specifically, the mutual receptions.

Mutual reception shows, as it were, friendship or enmity between two planets in the chart. These planets will signify various tendencies within ourselves and in the outside world through which we move. How this friendship or enmity will work will depend on the exact condition of the planets concerned. An example will make this clearer. Suppose I have Mercury seriously debilitated in the fifth house of my chart. It would not be a surprise to hear that my schoolwork suffers because I am too busy enjoying myself. No doubt ideas about a cure for this leap to the reader's mind; but what is an effective solution for one person will completely fail with someone else.

Suppose now that in my chart Saturn is in strong mutual reception with this weak Mercury. It is as if Saturn were Mercury's friend and so were willing to help him out. If Saturn is in dignity – in Capricorn, perhaps, or Libra – we can appeal to the better side of Saturn: "Pull yourself together! Get some self-discipline!" Such an appeal, backed up by the imposition of some stricter time-planning (Saturn), would work wonders.

But if the Saturn were weak – in Cancer, perhaps – there is little sense of self-discipline to which we might appeal. We have to turn to Saturn's less pleasant side, its fearful nature: "If you don't do well, the consequences will be dire!" With a debilitated Saturn, this approach will prove far more effective.

We must be particularly aware of the dangers shown by negative mutual receptions: that is, receptions involving the debilities of detriment and fall. We can either have outright negative reception, a reception that is positive in one direction and negative in the other, or even receptions that have a mixture of positive and negative in either or both directions.

This too makes little sense in the abstract, but becomes clearer with examples. Imagine that the rulers of the fifth house (fun) and tenth house (career) in my chart are in each other's detriment. They harm each other. My desire for fun ruins my career, while my career – because my concentration on amusement has left me in a dead-end job, working long hours for little money – spoils my fun.

For positive and negative reception in different directions, imagine the ruler of the ninth house (study) is in the exaltation of the ruler of my tenth (career), but the ruler of the tenth is in the detriment of the ruler of the ninth. I study well to get ahead in my job, but in my job I am so busy that I am unable to maintain my studies.

Or we might have mutual reception that is both positive and negative in either or both directions. This occurs most often with Mars in Cancer or Venus in Virgo, as they both have dignity by triplicity in these signs, yet are also in their fall. Perhaps the lord of the tenth house receives the lord of the Ascendant into both its triplicity and fall: I like my job, but it ruins my health.

Studying the web of receptions in the chart, always considering the accidental and essential strengths of the planets involved, gives us, then, a means by which we may tinker with the life – a method of finding a way of escape from the oncoming steam-roller. By manipulating the receptions we can stop banging our heads on whatever wall it is that is blocking our path and find a way of walking round it. Rather than confronting the issue head-on, in a way that is often counter-productive, we are locating a planet in the chart that is 'friends' with the offending planet, and who can have a quiet word with him on our behalf.

The intervention – the encouragement to our selected planet to go and have this quiet word with his chum – can be performed in many different ways, not only by shouting "Pull your socks up!" or "The bogeymen will get you!" as was suggested in the example with Saturn above. We have our full range of resources: alterations in the diet; use of precious stones; imaging; and so forth. Which of these it is appropriate to use will be indicated by the overall nature of the nativity with which we are dealing. The secret here is that each planet covers an infinitude of different levels, and we do not need to work only on the level at which it is manifesting. We can approach it in some unexpected way: instead of shouting advice, we add something to the diet, for instance. Such an approach stops us being drawn into the game-plan dictated by the offending planet, enabling us rather to choose the set of rules best suited for the job in hand.

What we are doing is akin to grand strategy. If our enemy has a powerful army, we do not allow ourselves to be drawn into battle by land. We seduce his allies to our cause and use their influence to persuade him to make peace. So with the chart and the mutual receptions within it.

This assumes that there is a suitable mutual reception in the nativity for us to use. Often there is not. We can, however, still do much by working with mutual receptions across charts – usually the charts of marriage partners, but also valuably with charts for people who work together, in order to enable them to bring the best out of each other. We might find it impossible to strengthen a weak planet in one person's birth-chart; but if that planet is in mutual reception with something suitable in the partner's chart, we can work on emphasising that quality in the partner. Maybe the afflicted ruler of my third house is in helpful mutual reception with the strongly placed ruler of your third: OK, you make the phone-calls, while I'll get on with something else.

These chains of reception often have much to do with the reasons for which people choose to live together in the first place, just as negative mutual receptions across the charts are usually at the root of recurrent conflicts. In days and places when astrologers had an important voice in the selection of marriage partners, the ways in which the charts received each other was given the strongest consideration. Its importance for the future of any relationship and the capacity of two people to grow together cannot be overstated.

In modern astrology there is still talk of mutual reception by sign, although little practical use is made of it. It is to be lamented that the study of dignity and reception, which is the very heart of traditional astrology, has been almost entirely abandoned, for it is the most powerful tool that we possess for making positive interventions in the life – whether this life be our own or our client's.

THE TRIAL

A new client phoned, asking a horary question about an impending court case. She was involved in a dispute with an ex-employer. He had issued a writ to force her to leave the flat in which she was living; she had issued a counter-claim to recover unpaid wages that he owed her. She wanted to know if she would win the case, if she would recover some or all of the debt, and if she would have to leave the flat.

The chart, as ever, was cast for the time of the question. As querent, she is given the Ascendant. At 26 Taurus, Caput Algol, the Medusa's Head, the most malefic of the fixed stars is rising. Algol is not among the brightest of stars, yet its power at least matches even that of brilliant first magnitude stars such as Regulus or Aldebaran. This is probably because

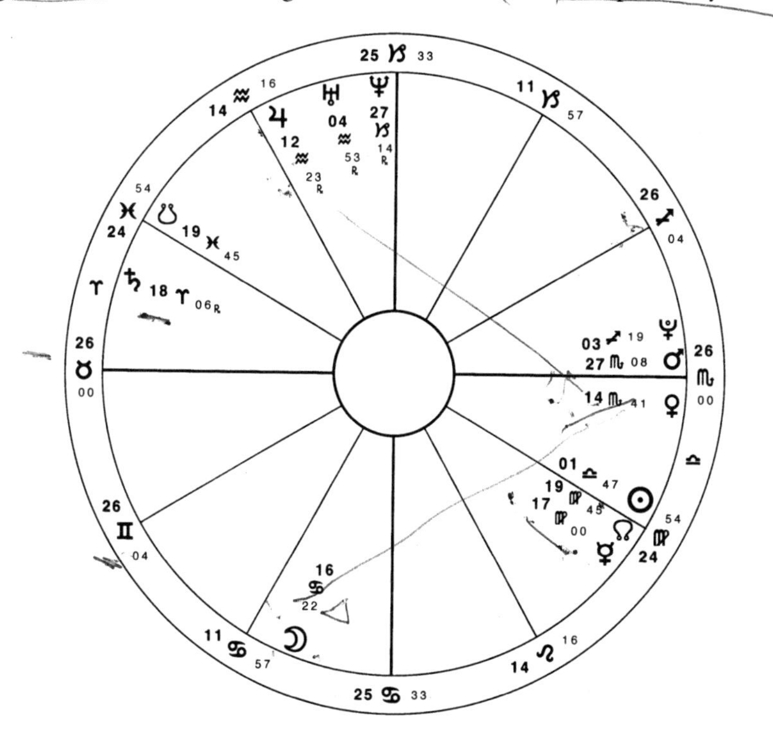

Chart 1. Will I win? September 24th 1997. 8.36 pm BST. London.

it is strongly variable – a double-star system which darkens as one of its elements passes in front of the other, giving the impression, and therefore the astrological consequences, of a permanent eclipse (astronomers know it as an *eclipsing binary*). This is an immediate testimony that proceeding to court is unwise: she is cutting her own throat.

In a court case, or any battle, we look at the strengths of the respective significators. Hers is Venus, ruler of the Ascendant, her opponent's Mars, Lord of the seventh, the house of open enemies. Venus is dreadfully weak: she has dignity by term, but is in the sign of her detriment, a cadent house, and is disposited by Mars, her enemy. She is also conjunct another of the more malefic fixed stars, the South Scale. Things do not look good.

Mars, by contrast, could scarcely be stronger. He is in his own sign and just inside his own house – an angular house at that. He is untouchable, especially as the sign in question is Scorpio, a fixed sign, rather than Aries: he cannot be shaken.

But a court case is not solely a trial of strength. Were it a wrestling match, the stronger would invariably win; but a court case is decided by a judge, not by brute force. This can lead to eccentric and unexpected decisions in favour of the weaker party. The judge is shown by the tenth house and its ruler, in this case Saturn.

Lilly says that if Saturn is his significator, the judge 'will not judge as he ought'. All the more so here, as Saturn is seriously debilitated: peregrine, retrograde and in the sign of its fall. His decision is plain: disposited by Mars and in no dignity of Venus, he will judge in favour of our querent's opponent.

We look to the ruler of the fourth house to show 'the end of the matter': the verdict. Here, it is the Moon. She separates from aspect to Venus, which is in itself an unfortunate testimony, then picks up Mercury (Lord of the second, i.e. the querent's money), carries it to Saturn, the court, and finishes by trining Mars. Judgement will go to the opponent. In this case, the Lord of the fourth is also relevant as significator of the contested flat: again, it ends up with the opponent.

The money that the querent is claiming is shown by the eighth house, the second from the seventh and so the house of the enemy's money. This is ruled by Jupiter. Jupiter and Venus, the querent, are mutually separating from aspect, as Jupiter is retrograde: the querent and the money are moving farther apart. She will not get any of it. All in all, prospects are dismal.

The trial itself

The chart set for the time and place of the court hearing paints the same picture. This is an event chart, not a horary, so her opponent, as the immediate instigator of proceedings, is shown by the first house. Its ruler, Jupiter, is much less strongly shown than in the horary; but it matters little how strong or weak it is, for Mercury, ruler of the seventh, and so significator of the enemy – in this chart our client – is combust. All its power is destroyed and she can have no hope of winning.

The judge is still shown by the Lord of the tenth: Venus. Venus is peregrine, again casting doubt on the soundness of the verdict, and is immediately inside the Ascendant. A decision against our client could scarcely be more strongly shown.

Saturn afflicting the fourth cusp repeats the suggestion of the horary that justice will not be done, while Mars, ruler of the fourth and so significator of the verdict is also tucked just inside the Ascendant. As a final testimony, Fortuna (17 Virgo 04), which will show the 'treasure' of the

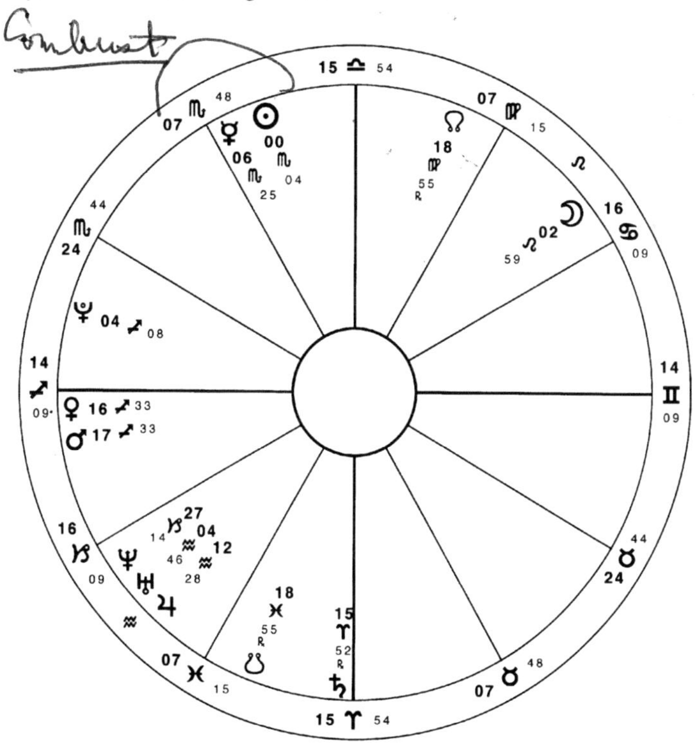

Chart 2. Court case: event chart. Data withheld.

instigator of the action, is exactly conjunct the benefic North Node in the ninth, the house of the law. He will win.

A Third Chart

The client was not happy. She had first contacted another astrologer, who does not practice horary and so had referred her to me. 'Can you use the time at which I asked him the question?' she asked. 'Perhaps it will give a different answer.' The important moment is that at which the astrologer understands the question – so its having been asked to someone else beforehand is of no more consequence than the querent's turning it over in her own head, but out of curiosity I set the chart for her first attempt to ask it.

The early degree rising suggests that the question may be premature, as this astrologer's inability to answer it shows that it was. But even with different planets signifying the querent and her enemy, we have the same indication: the opponent's planet is once again in its own sign, just inside

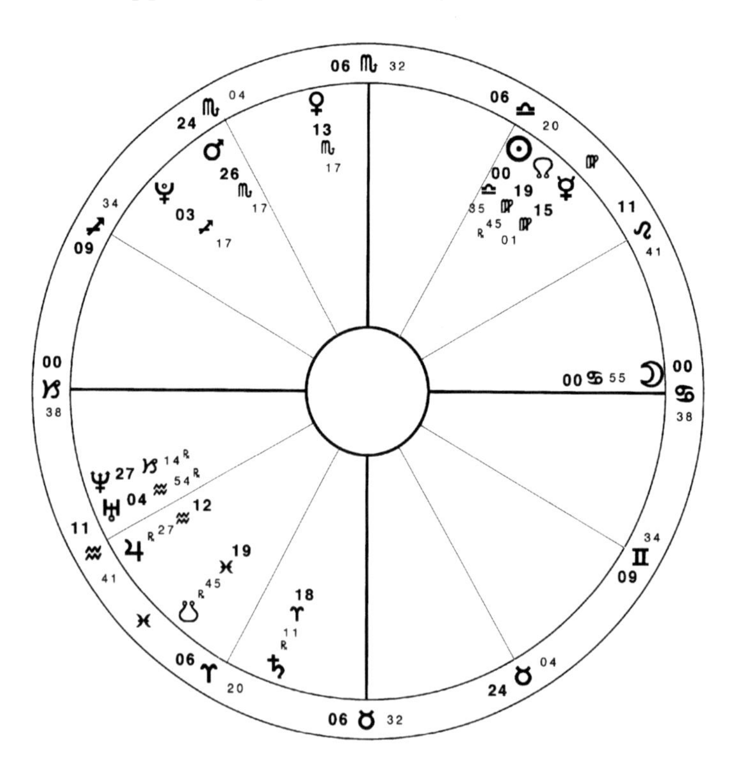

Chart 3. Will I win? Data withheld.

his own house. The querent's planet, in this case Saturn, is again very weak. She will not win.

The court looks rather better in this chart, being shown by a dignified Mars. Saturn and Mars make no contact, while the Moon – the opponent – makes its last aspect to Mars: he will win. The ruler of the fourth is in detriment in the tenth house: once more, we have the idea of a wrong verdict. (The querent says that this is but one of a long string of court cases her opponent is facing, but he seems untouchable, however much evidence is laid against him.)

Finally, Fortuna (0 Libra 58) is combust and its dispositor in its detriment. This chart offers no better prospect than the others. The querent will lose.

This raises the issue of whether one can ask the same question twice. We are clearly told by the authorities that we cannot; but in practice it does often happen, most frequently with 'Is there any hope for this relationship?' questions, where the querent is torn between two difficult choices, neither of which she wishes to take.

There are those who say that the Big Questions are unanswerable by horary for just this reason: as soon as a general election is announced, for example, astrologers throughout the land are asking questions that have identical wording. But only the dilettante can say that they are asking the same question, for implicit in astrology – the very root of astrology – is the principle that what happens at any moment is unique to that moment. So 'Will Labour win?' asked by me now is not the same question as 'Will Labour win?' asked by my colleague in two minutes' time.

In the same way, 'Shall I leave him?' asked today is not the same question as 'Shall I leave him?' asked in two weeks' time. They do have an organic relationship: if the issue under discussion is pictured as a worm, the two questions can be seen as cross-sections of that same worm, in much the same way that progressed charts give us cross-sections of a life. Or in the same way that the birth-charts of different family members give us cross-sections of the being that is that family. Indeed, the charts for repeated questions do tend to relate in the same way as the birth-charts of family members. The echoed key to judgement in these charts – different Ascendant axes, but Lord of the seventh in an identical position in both charts – is just the kind of theme we would expect to find repeated in related nativities. These familial-type recurrences – notably different planets in the same positions or the same planets in different aspects – occur in both similar questions asked by the same person and also questions on the Big Issues asked by different people. Obviously the problem of 'Shall I leave him?' has an organic life within the heart of the querent; it seems too that the question 'Will Labour win?' is an organic entity in itself, so any cross-section taken through it will reveal a congruent internal structure, whether that cross-section be of its manifestation in me or in my neighbour.

When we are pulling the petals off flowers asking 'She loves me, she loves me not,' we will pick another flower and try again if we don't get the answer we want. Treating horary in this way is foolish: if we do not believe the first answer, there is no reason why we should believe the second. But there are many circumstances in which a second question may validly be asked.

We may want a second opinion, just as from a doctor: and if I have measles when I see this doctor today, I will still have measles when I see that doctor tomorrow: the symptoms in the horaries will be the same. We may want more reassurance or clarification, and there is no reason why we cannot go back for another look – truth is sturdy enough a beast not to run and hide as soon as we catch a glimpse of it. We may have legitimate doubts about the abilities of the astrologer who judged our question. Any of these reasons will yield a valid second – or third or fourth – question.

What the stricture is aimed at is those who have nothing better to do than to 'test' the astrologer; avoiding these, indeed, seems Bonatti's main concern in enumerating his considerations before judgement. Clearly, horary can never be used to test the astrologer, for, regardless of what question is actually spoken, the real question then is always 'Is this astrologer competent?' and we might as well ask 'Are you lying?' But a question asked with sincerity will always be valid, no matter how often those same words have been used before.

THE DEFAULT OPTION

One of the great virtues of astrology is that it allows us a dispassionate view of whatever situation we are examining. So when I look at a chart cast for my question "Does Susie love me?" I can set aside my assumptions that no woman can resist me and see quite clearly that Susie is the exception who can. Simple enough in theory, and, indeed, simple enough most of the time in practice.

Most of the time, then, astrology enables us to keep our presuppositions locked securely out of the way of our judgement, and we are well advised to cooperate with astrology by striving to our utmost to avoid importing our prejudices. Every romance I have ever had might have been a disaster, but if my client's question is "Is there a future in this relationship?" I must avoid judging the chart by the yellow light of my own jaundiced view and see instead only what is before me.

On occasion, however, we need to be aware that we are judging within our own, necessarily partial, view of the situation. This concerns what we might call 'the default option': that is, what happens when the chart shows nothing happening? In most questions, the default option is inbuilt: "Will I marry Susie?" and nothing in the chart, the answer is No; as with "Will I get the job?" or "Will I win the lottery?" But if the client phones on the morning of his wedding, just to make sure everything is on line, asking "Will I marry Susie?" our chart with nothing in it would give the default answer Yes. If no disruption is shown, all will proceed as planned.

The problem is really one of accurate phrasing of the question. Our bride-groom-to-be does not mean "Will I marry Susie?" but "Is anything going to go horribly wrong?" so our default option is as before: No – nothing happening in the chart, nothing happening to disrupt the wedding. As, however, it is unlikely that our clientele consists entirely of schoolteachers, it is unlikely that the questions we receive will all be phrased with pedantic nicety; the consequences can be confusing. I was asked whether the election in Pakistan would take place. The picture I had of the situation there was one of extreme volatility, so my assumption was that it very probably would not. The chart showed nothing happening, so I judged No. Had my understanding been clearer, I would have known that the chance of the Pakistani election not taking place was little greater than that of the elections in Britain or America failing to happen, so the chart with nothing happening should have given the answer Yes: no disruption is shown, so nothing untoward will happen and all will proceed as planned. The astrologer's failure lay not in the judgment of the chart, but in his inability, through the fog of his own assumptions, to elicit the necessary degree of clarity in the client's question.

This is something of which we must be aware. Particularly, it seems, in questions of conception or relationships, it is sometimes necessary to beat one's clients violently to make them give up simple pieces of information which are vital for the correct understanding of the chart. The perennial difficulty with judging one's own charts has much to do with this, though here the major problem is of too much, rather than too little information. The Saturn on the Ascendant which destroys our client's question is easily explained away as Granny coming to tea if it occurs in our own chart, allowing us still to live in hope of whatever unlikely gratification we have decided we deserve.

We must also be aware of this question of defaults when considering separating aspects. The separating aspect in the chart, having already happened, usually shows an event that has already happened at the time the question is asked. But it can show that all will proceed according to plan. It is as if the aspect has been put in motion, and as there will be no obstruction, so far as the chart is concerned the event is already done.

This client had booked a long trip overseas. To make this trip, she had to let her house; this she had arranged, but she was worried that the builders who were making improvements to it would not finish before she left, the deal would fall through and she would have to cancel or postpone her journey. The relevant aspects in this chart are all separating. If we stick to a literal interpretation of the rules, this makes no sense; but if we invite Nicholas Culpeper's esteemed brethren, Dr. Reason and Dr. Experience, to assist in our judgement it falls into place.

With Cancer rising, the querent is shown by the Moon. She is separating from a conjunction with Saturn, Lord of the ninth house of long journeys. Obviously, the querent has not made the journey already, or the question should not be asked; we must judge that it will proceed as

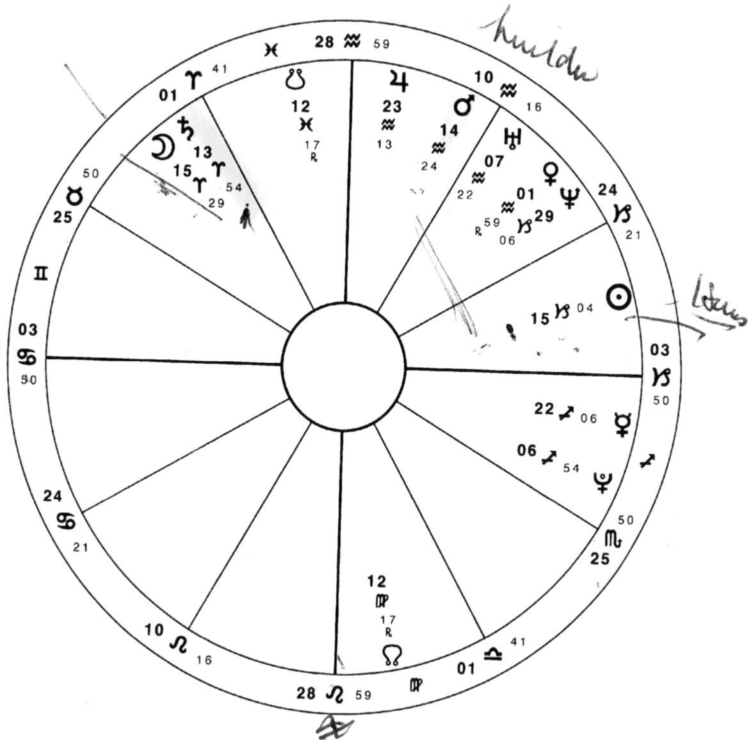

Chart 4. Will the builders finish? January 5th 1998. 3.05 pm GMT. London.

planned. The ninth house is afflicted by the presence of Mars, which rules the sixth house of slaves and hence shows the builders. With both Saturn and the Moon (the journey and the querent) ruled by Mars, the importance of the builders is apparent; that Mars is itself ruled by Saturn, Lord of the ninth, indicates that the builders are there only because of the journey – the work would not otherwise have been ordered. The Moon separates from easy aspect to Mars: again, the querent and the builders have not yet parted company, but we must judge that this parting is under way. The affliction to the ninth still holds, but it will not come from the builders: the placement of both Moon and Saturn in the eleventh house of hopes and wishes, also ruled by Mars, shows the particular source of the problem – one that would involve an unnecessary digression to deal with here. Finally, we have the business of letting the house. As ruler of the seventh, Saturn would also show the tenants, while the Sun, Lord of the fourth, is the house. Again we have separating aspects, both between the house and the tenants and between the querent and the house (the Moon has just translated light between tenants and house, bringing them together). Again, these events have not yet happened, but must be judged as all proceeding according to plan. The mutual reception between the house and the builders is a cause for concern: with the house being received into its detriment by Mars which is itself peregrine, we must doubt their competence. But they will at least be finished.

And that is what happened: the builders finished their work, the tenants moved in and the querent jetted off on holiday. We must also note here two important points about mutual reception, as the receptions between the various players in this drama are of significance in determining the outcome of her trip: mutual reception can strengthen a planet only insofar as the planet with which it has this reception has any strength to give it; if the planet itself is too weak, it cannot be helped much even by the strongest of partners.

The first of these is common sense: if neither planet has any strength, this strength is not going to be conjured up from nowhere just because they receive each other. Two tramps may like each other no end (mutual reception), but they are still tramps. Thus with planets that are peregrine, or, in the traditional term, 'like homeless wanderers'; all the more so if the planets in question are in detriment or fall: my best friend may plead, "I've lost all my money – help me out," but if I am in the same boat, as debilitated as he, my pockets too are empty and for all my friendly feelings I can be of no assistance.

A weak planet cannot give substantive help; nor can it receive it. This is known as 'reflecting light': the light is given through the reception, but if the planet is in detriment or fall it cannot hold onto it, rather like a sick man who is fed but cannot keep anything down. Some of the authorities state that planets thus debilitated cannot be in mutual reception at all; this seems an exaggeration, as even in that state, Dr. Experience tells me, it does take the edge off things a little – as with my impoverished friend and I: we cannot offer each other material help, but we can give sympathy. This may not pay the bills, but it is better than nothing.

NEW OLD TECHNIQUES

When first we start studying traditional astrology, the grumpy ghost of William Lilly appears and snatches away all our favourite toys. Outer planets, asteroids, minor aspects: all are gone, locked securely in the toy-cupboard, not to be released no matter how well we may behave. That sesquiquadrate between Chiron and Transpluto that seemed the lynchpin of my personality is gone forever.

But while we kneel weeping in the playroom, lamenting the cruelty of fate and our sad loss, a good fairy appears, bearing toys brighter and better by far than those of which we are bereft. Best and brightest of all these new toys are antiscia and Arabian Parts: fun for all the family, and – best of all – they really work! Yet even among the journeymen in the Apprentice's workshop, there are few, familiar with these techniques though they are, who use them as often as they might. The Master can often be heard, grumbling over his pipe at the end of a tiring day, complaining that "If only they'd look at the antiscia, their judgements would be so much more reliable," or while taking delivery of another load of Arabian Parts from that nice Moorish gentleman with the aromatic tobacco, "It's all very well, but no one apart from me ever uses most of these, so they'll just go rusty at the back of the workshop."

Arabian Parts are wonderfully incisive. If horary is to astrology what surgery is to medicine, Arabian Parts enable us to perform keyhole surgery. We may consider, for example, the pages that the ancients devote to the diagnosis of disease: a quick look at the Part of Sickness and its dispositor will often provide much the same information with only a fraction of the time and effort.

There are Arabian Parts for almost everything, from Assassination to Cucumbers – the point of a Part of Cucumbers being to determine a good time to sow or sell one's crop, or to find how the cucumber harvest will be affected by this eclipse/comet/lunation/ingress or whatever. Once you have the hang of the technique, it is possible to make up your own parts in the unlikely event of the one you want not already existing. We must, however, cast extreme doubt on the bastard technique of American Parts, constructed using Uranus, Neptune and Pluto.

If we are turning the chart, we must turn the Parts: if I am judging a horary for my own question about my daughter's career, there is no point in considering the radical Part of Vocation – it is her vocation, not mine, that is important.

The idea of antiscia is a strange one. There is something thoroughly unconvincing about the suggestion that we can find what amounts to an alternative placement of a planet by reflecting its position in a line from 0 Cancer to 0 Capricorn. It sounds like something that has been pulled out of the air. But it works. A judgement based on an antiscion is as reliable as any other, as this example shows.

Antiscion is literally 'shadow', and a planet's antiscion is best considered as an alternative placement of that planet. It takes only seconds to glance around a chart to check if anything interesting is happening to the antiscion of any of the relevant planets, yet these few seconds can be vital: even the rosiest judgement, for example, can be jeopardised by the antiscion of one of the significators falling on a malefic.

Mr. Sharif, the Prime Minister of Pakistan, had been accused of corruption. His case was in process of coming to court. My client described the situation as one of open war between Sharif and the Chief Justice of the Supreme Court. If the case were heard, Sharif would be disqualified, so he was intent on preventing this happening. If he were disqualified, he would be replaced by his brother (the deputy Sharif). The question asked was "Who will win: Sharif or the judges?"

The client was Pakistani, so Sharif is given the tenth house of rulers (had I asked the question myself, I would have taken the tenth from the ninth, as he would then be the ruler of a foreign country). He is signified by Saturn, which is in a dreadful state, in its fall, peregrine and retrograde. It is, however, in the house of its joy, so all is not lost. It is perhaps as if he has been pushed back to the final citadel, or perhaps that, however parlous his position, he is still the holder of power.

In this instance the judges are his open enemies, so are given the seventh house from the tenth: the fourth. So they are signified by the Moon, which is in the double-bodied sign of Virgo, showing that there is more than one of them. The Moon has dignity by triplicity, making it far stronger than Saturn: the judges have the upper hand. This is emphasised by the position of the Part of Resignation and Dismissal (Saturn+Jupiter-Sun), which falls at 24.55 Cancer, just inside the fourth house, showing that dismissal is within the judges' power.

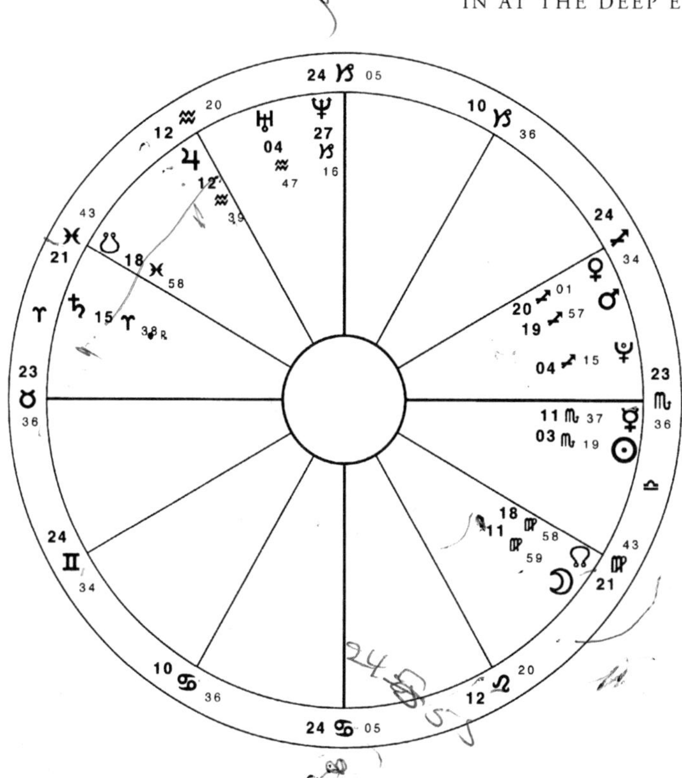

Chart 5. Who will win? October 26th 1997. 6.24 pm BST. London.

But however weak Saturn might be, it is not getting any weaker, nor is it about to change its state in any way. It makes no major aspect with either the Moon or the Part of Dismissal. There is no indication of anything happening to Sharif. His brother is shown by Jupiter, ruler of Pisces, the sign on the cusp of the third house (brothers) from the tenth. This too is doing nothing of note. Jupiter and Saturn are both applying to a sextile, with strong mutual reception, but this would hardly show one replacing the other – and even if it did, the aspect is prohibited by the Sun.

There is no sign of Sharif falling – so what is going to happen? This is shown by the antiscion of the main significator. The antiscion of Saturn falls at 14.22 Virgo: the Moon applies immediately to conjunct it. In a question of 'Law-suit or Controversie betwixt two' Lilly states that if the two main significators are conjunct, 'the parties will easily of themselves accord, and compose all differences without mediation of any, or with a little entreaty.' This was the judgement. The client, who was well-placed to know, assured me that it was impossible: there could be no conciliation between them. A week later, he phoned to say that it had happened: Lilly's experience and the technique of antiscion had proved correct.

Antiscia.

The antiscion should be regarded as an alternative placement of the planet (or Arabian Part) concerned. It is just as if the planet were in that place as well as in its own. Conjunction or opposition to an antiscion (opposition to antiscion being a simpler way of regarding contrantiscion, it being always worth avoiding the unnecessary proliferation of technical terms) will work just as well as conjunction or opposition to the original position of the planet. Other aspects are not so strong, and should be regarded as supporting testimony only. All other things being equal, a horary chart can be judged on an unsupported antiscial conjunction or opposition, as we have done on many occasions with verifiable results. Conjunction and opposition with antiscia carry exactly the same meanings as normally, except that the antiscia (if the context supports this) tend to carry a sense of the covert, presumably through their, literal, nature as shadows.

The common stumbling-block is movement. Do not try to move antiscia: you will tie your brain in knots! Yes, as the planets move forward their antiscia theoretically move backwards, and *vice versa* if they happen to be retrograde; but working with this is a quick way to the insane asylum: in the workshop we always make sure a bucket of water is kept handy to throw over any young apprentices who try it. Take it as a rule: things apply to antiscia, antiscia do not apply to things.

The idea that a planet can somehow be conjunct another planet that is on the other side of the chart, is apt to strike the student as bizarre. Even when experience has shown that it works accurately and reliably, it is still hard to understand why.

We may attempt to clarify the matter by imagining the cosmos as a watermelon. The Earth is right in the middle, and around the longest circumference of the watermelon has been drawn a band an inch or two wide, on which are inscribed the signs of the zodiac. This band has been marked off into 360 degrees.

We now take a knife and start paring slices from one end of the watermelon, perpendicular to the axis that runs through the the centre of the melon, which axis passes through the zodiac we have drawn at 0 Cancer and 0 Capricorn. If we continue to cut slices, each one degree thick, we shall end up with 180 slices. Each of these slices will have a degree marked on either side; these degrees will each be the other's antiscion. (Yes, you can try this one at home!)

If we take another watermelon and repeat the exercise, this time cutting slices perpendicular to an axis passing through 0 Aries and 0 Libra, we again end up with 180 slices. Each of these will have two degrees marked on it, each of which will be the other's contrantiscion.

What, apart from a tasty snack, does this give us? It does point the connection between the degrees in question. It is easy to forget that our astrological degrees are degrees of longitude, so, as on a terrestrial globe, they run all the way up and down; they are not just small segments of the line that is the zodiac. If we see a planet in, say, the third degree of Taurus and, keeping the same relative distance from the solstice points, follow it all the way up to the top and down the other side, we arrive at its antiscion. Planets in the same degree of celestial longitude are in conjunction; antiscion just takes this to an extreme.

This must cast doubt on the common claim, made particularly strongly by Placidus, but also by other authorities, that planets must be in aspect by latitude as well as longitude for the aspect to be effective. We know that conjunctions and oppositions by antiscion are effective; yet as we have seen, to make these we have to travel all the way up to the top and down the other side. It is, then, difficult to see why a few degrees up or down should make any significant difference.

The pattern in which antiscia and contrantiscia combine signs is worthy of note. Antiscia match hot/dry with cold/dry and hot/moist with cold/moist. There are some grounds for understanding between these elemental combinations. Contrantiscia combine utterly incompatible signs – even more so than does opposition – matching hot/dry with cold/moist and cold/dry with hot/moist. The significance of this becomes apparent when we consider to which axis each of these techniques pertains.

Antiscia can be regarded as resolving themselves along the 0 Cancer/0 Capricorn axis: the axis of the solstices, of Regulus and Fomalhaut (although they are no longer in these places). This is the vertical post of the cross. It is the initial thrust of creation, from hot/dry to cold/dry. From the cold and dry came moisture (cold/moist), which when warmed became air (hot/moist). Hence the two groups, fire and earth, air and water, into which antiscia divide the signs.

Contrantiscia give the horizontal bar of the cross. The axis of birth and death, Aldebaran to Antares. Innocence and Experience; the soul hungry for experience and the soul that longs to return to innocence, that has had enough and wants to go home. Hence the total incompatibility of the signs connected here, the tension that cannot endure, for what comes into life along this axis must go out of it again. The significance of these axes may well explain why antiscia and contrantiscia are so common in charts relating to death.

Arabian Parts:

Arabian Parts, if used correctly, can be invaluable. Alternatively, they can make a nonsense out of any chart. The first misapprehension that must be corrected is the idea of Parts 'helping' in any way. Time and again we read accounts of chart judgements where we are told 'the Part of (whatever) was sextile our main significator, but as it was in the twelfth house it could not be of any help', or similar. Parts are; they do not do. They are not bodies; as such, they have no light. What has no light has no power to act.

The planet dispositing the Part signifies whatever that Part represents. So if the Part of Death falls in Aries, Mars signifies death. Main significator conjuncts Mars: strong testimony for death. Part of Turnips falls in Cancer, Moon signifies turnips. Moon conjuncts Jupiter: my turnip crop does splendidly.

The common practice in astrological software is to get the judgement of Parts exactly the wrong way round. The software sets the chart and provides a list of all the Arabian Parts that fall conjunct planets in it. But if my question is "Will she marry me?" the fact that the Part of Turnips happens to be conjunct Venus is irrelevant. Correct practice is not to scan a list of Parts to see which ones are drawn into the chart, but to calculate the Part for whatever it is that interests us and then see what is happening to it.

The Parts that I find most often useful are these. An R shows that the final two terms reverse in night charts. I am unconvinced by the arguments for reversing Fortuna and the Parts based on it, and have achieved good results by not doing so.

It is common to reverse the formula for the Part of Marriage in order to obtain a Part of Divorce. This is wrong, as it implies that divorce is either the opposite of or the complement to marriage, neither of which is so. I suggest instead using the traditional ruler of divorce, Mars, in the formula Asc+Desc-Mars. This both makes sense and works.

Fortuna:	Asc + ☽ - ☉
Spirit:	Asc + ☉ - ☽
Love:	Asc + Spirit - Fortuna
Despair:	Asc + Fortuna - Spirit
Captivity:	Asc + Fortuna - ♄
Victory:	Asc + ♃ - Spirit
Courage:	Asc + Fortuna - ♂
Marriage:	Asc + Desc - ♀
Marriage of Women:	Asc + ♄ - ♀ R for men; both R by night
Vocation:	MC + ☽ - ☉
Sickness/Violence:	Asc + ♂ - ♄ R
Surgery/Cure:	Asc + ♄ - ♂ R
Death:	Asc + 8th cusp - ☽ or 8th cusp + ♄ - ☽
Fame:	Asc + ♃ - ☉ R
Marriage Partner:	Asc + Desc - Lord 7th
Divorce:	Asc + Desc - ♂
Servants:	☿ + ♄ - ♃
Employers:	☽+ ♄- ♃
Dismissal/Resignation	♄ + ♃ - ☉

BORN TO BE BAD

There was a commotion in the workshop the other day. A selection of planets from a chart on which the Master was working had been allowed into the yard to stretch their legs while the house-cusps were being prepared. Usually Diggory and myself are there to keep order when this happens, but on this occasion, just at the time when Diggory had had to slip behind the stables for a moment or two, I had to deal with an urgent message. This shows what a two-edged sword the progress to which we here are so open really is, as in the old days the carrier pigeons would arrive in the pigeon-loft to be dealt with in due course; but now we are forever subject to the strident tones of a new kind of pigeon, bred specially to come straight to the individual for whom the message is intended, wherever he might be or however important might be the task with which he is currently involved.

So, just as Diggory had disappeared, my attention was distracted by the raucous arrival of one of these new 'mobile' pigeons. The message was nothing of any great significance, merely one of those tiresome enquiries as to the condition of our windows and our willingness to take part in a local promotion; but by the time I was once again able to fully devote myself to my duties, turmoil had erupted among the planets whose exercise I was supposed to be supervising.

The trouble seemed to have been started by Mercury. I had seen him hanging around with Mars, and this association usually leaves him rather too sharp for his own good. The two of them had had a falling out, on which Mercury had begun running around Mars and Saturn at great speed, taunting them with cries of "You're malefics, you are!" Saturn, who had a station approaching and so was feeling rather vulnerable, had burst into tears; Mars, despite some initial trouble catching him, had finally managed to corner the miscreant and, by the time I was able to intervene, was dealing with him in much the manner that we might expect.

I grabbed them all by the scruff of the neck and hauled them off to the Master. He gave them a good talking-to, but Mercury was unrepentant, repeating constantly, "They're malefics and I hate them!" The Master then pulled out his pipe and entered upon a detailed explanation of just what causes a planet to be a malefic, by the end of which even Mercury was clearly feeling rather less smug.

In the land of the tradition we do rightly pride ourselves that our astrology has malefics and benefics. Ours is not the candy-floss world of the moderns; we do face the fact that all in our garden is not lovely. But for all that, we must avoid the careless use of the terms 'malefic' and 'benefic' that turns them into labels that obscure the truth. At worst, 'malefic' is debased almost into an epithet of abuse for any planet that we don't like, not so dissimilar to 'fascist' or 'pinko'.

Let us consider an example. This horary was set for a question about a burglary. Someone had broken into the querent's flat, scattered her possessions around, but not taken anything. Despite police assurances that this was by no means uncommon, our querent was worried that the intruder might have been a stalker. So she asked, "Did the burglar know me?"

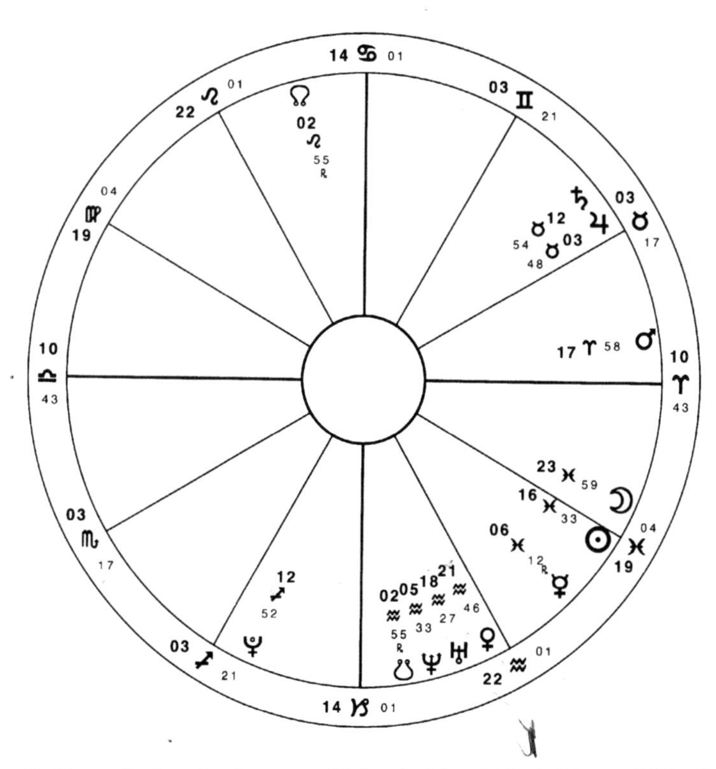

Chart 6. Does the burglar know me? March 6th 2000. 8.02 pm GMT. London.

The first candidate for a thief in a horary chart is a peregrine planet in an angle or the second house: there is none here. The next choice is the ruler of the seventh, the house of thieves. This is Mars, which is in its own sign, Aries, and so is strongly essentially dignified. Lilly tells us that when a malefic is essentially dignified it behaves itself better, for it is 'like a noble soul that hath the enemy in his clutches, but scornes to hurt him'. (Note that this applies only to *essentially* dignified planets. Malefics that are essentially weak but accidentally strong are just better able to work their malice.) This is our answer: the burglar has broken into the flat, but has disdained to steal anything. Our poor querent's possessions were not good enough for this 'noble soul'! In Aries, Mars exalts the Sun, so perhaps he was looking for gold, or at least something of rather higher value than what he found.

If the burglar had been a deranged admirer, we would expect to find his planet in dignities of Venus, the querent's significator. It is not. Indeed, in any relationship question one of the strongest indicators of lack of interest from the quesited is finding his planet in his own house, as it is here. The chart is clear: our querent has nothing to worry about. All we have here is a very fussy thief.

This example shows a well-behaved malefic. Another chart concerned someone who had suffered a severe allergic reaction to a soft drink. The drink was signified by Venus, the natural ruler of sweets, which was in the sign of its fall. Here we have the opposite situation: an essentially debilitated benefic. Again, the chart accurately describes the circumstances: where before we saw an essentially malefic thing – a thief – disdaining to do evil, here we had an essentially benefic thing – a soft drink – doing harm.

We know that not only Mars and Saturn are malefics. If the context is appropriate, any planet ruling the sixth, eighth, twelfth or even fourth house can temporarily acquire malefic status. If the Lord of the eighth is out to get you, it is no good reminding it that it is a benefic: it will kill you none the less – a point which gives a salutary reminder of how our conceptions of good and bad are limited by our petty concerns.

Any planet prohibiting an aspect in horary is likely to be seen as a malefic, no matter which planet it might be, simply because it prevents us attaining whatever it is that we think we want.

We must return to the underlying philosophy. The universe is created by God; God is infinitely good; we cannot, therefore, accept the idea that

Mars or Saturn is intrinsically evil. They are not evil; they are just inconvenient. We can follow the primrose path of Venus to our eternal destruction and still regard her as a benefic, because we think we are enjoying the ride. But even a fraction too much Mars in our curry strikes us as deeply unpleasant, to say nothing of the sharp dose of Mars that keeps us from getting tangled in the briars at the side of our road.

When Mars or Saturn is strongly dignified it is behaving more like its true self. It is not the essential nature of either planet to go about making a nuisance of itself because it is bored: their true natures are as necessary and, above that, as desirable as those of any other planet. It must be remembered that when, for instance, Mars moves from Aries to Taurus, from strong dignity to strong debility, the planet itself has not changed one jot; all that has changed is the starry background against which we see it: the context, as it were. Mars and Saturn indicate things which, for the most part, we do not happen to like. They are indeed malefic within the bounds of our immediate concerns, but for a full understanding of the nature of these planets we must attempt to see them in a wider context. Iamblichus explains: 'I am of opinion that what appears to us to be an accurate definition of justice does not also appear to be so to the Gods. For we, looking to that which is most brief, direct our attention to things present, and to this momentary life, and the manner in which it subsists. But the powers that are superior to us know the whole life of the soul.'

Ficino says: 'No parts of (the cosmos) can be inimical among themselves in any way. For fire does not flee water out of hatred of water, but out of love for itself, lest it be extinguished by the coldness of the water. Nor does water extinguish fire out of hatred of fire; but it is led by a certain desire for multiplying its own cold to create water like itself out of the body of the fire... The lamb does not hate the life and form of the wolf, but is fearful of its own destruction, which is occasioned by the wolf. Nor does the wolf kill and devour the lamb out of hatred for him, but out of love for himself.' This does not prevent the lamb perceiving the wolf as a malefic; but the more content the wolf is within himself (that is, the more essential dignity he has), which contentment is best occasioned by being replete, the more the lamb can sing and dance in front of his nose without fear of molestation. It is only when the wolf is feeling desperate, perceiving himself through the unpleasant sensation of an empty stomach as being unloved by the universe (i.e. in detriment or fall), that the lamb has something to worry about.

NEPTUNIA REPLIES...

– a word from our sensitive seer

Dear Neptunia, Who can I turn to but you? I'm about to get married; but I've found that my boyfriend keeps treating squares as if they were trines, when they fall in signs of long ascension. I've asked him to stop smoking that stuff, but he swears that has nothing to do with it. How can I ever trust a man who does this sort of thing? Please help me,

Yours in desperation, *Tracey*

Dear Tracey, You've found out just in time! Lots of men have these habits, but this doesn't mean you have to go along with them. It can lead only to tears.

I expect your boyfriend (*ex*-boyfriend by now, I hope) picked up this idea from William Lilly. Now, I'm as big a fan of Lilly as anyone – many a night I've spent hanging around outside the most fashionable nightspots for a glimpse of his impish good looks – but we should always bear in mind Mr. Culpeper's dictum, 'Let every one, that desires to be called by the name of Artist, have his wits in his Head (for that's the place ordained for them) and not in his books.'

Even at the comparatively early stage of his career at which he wrote *Christian Astrology*, Lilly had a depth of experience that few of us are likely ever to match. This helped make him a great astrologer; it did not make him infallible. If your boyfriend thinks it did, I suggest you subject his fourth house to some careful scrutiny. There are times when Lilly too seems to have been sampling the contents of Mr. Culpeper's herb-garden: this is one of them.

The claim is, that signs of long ascension (♋ to ♐) can be regarded as if they subtended more than 30 degrees. That is, we can regard 90 degrees in these signs as stretching as far as 120 degrees elsewhere, so a square can turn into a trine. Let us consider this.

First, if 90 degrees in signs of long ascension can be regarded as 120 degrees – an increase of one third – the 180 degrees from 0♋ to 0♑ must be taken as being 240. Lilly nowhere does this, for the obvious reason that it is absurd. Moon oppose Sun would no longer give a full Moon (except at 0♈/0♎). The North and South Nodes would no longer be opposite each other. The equinoxes would no longer be six months apart. And so on.

Second, and even more important, is the fact that this idea contradicts the basic principles on which the nature of aspects is based. A trine is a harmonious aspect because it brings together planets that are in the same element – earth to earth, fire to fire, etc. Squares are inharmonious because they bring together planets in inharmonious elements, mixing earth and water with fire and air. If we have an aspect between a planet in a fire sign and a planet in an earth sign, we can call it a square, a trine or a fried fish: it will still be inharmonious.

This is why aspects do not work unless the signs in which the planets fall are themselves in aspect: an astrological fact forgotten by modern astrologers. So, for example, 29♉ does not trine 1♎, not because there is some invisible barrier separating them, but because they are not in signs of the same element. When Neptunia was learning her astrology, more years ago than I care to remember, we were taught these basics before dealing with the intricacies of Pluto transits in our lunar returns.

We must not forget that Lilly was writing at a time when the empirical approach to knowledge was climbing to its present position of ascendancy. However incompatible this mode of thought may be with astrology, he could not help but be ensnared by it from time to time. This is one time.

If the aspect we find in the chart doesn't fit our idea of the situation, I would suggest that before we start changing the rules of the game, we devote some closer study to both the situation and the chart. Is your boyfriend quite sure he isn't imposing his preconceptions of what ought to be happening? Has he really explored every corner of the maze of receptions that tell so much about what is going on? Only then would he be in a position to tell the chart whether its squares ought to be trines.

You're better off without this one, Tracey. He may think he's trined to you now, but this sort of man can form a trine with anyone if the mood takes him. Cancel that wedding immediately!

Your caring, Neptunia

2

Let's Get This Straight

LET'S GET THIS STRAIGHT

Listen up and let's get these points sorted out once and for all. Our flying-squad of examiners will be calling on you in the near future to make sure you've been paying attention.

The Rule of Three:

Let us begin with the so-called 'rule of three': the idea that we must have at least three testimonies saying the same thing in order to make a judgement on a chart. We should be so lucky! This, like the extreme overemphasis given to the Considerations before Judgement and their elevation into 'strictures' is one more example of the Rule of Rules in modern horary: "Never give judgement if you can possibly avoid it."

Your Place or Mine?

We note with a familiar mixture of amusement and despair the propensity of astrologers, particularly it seems those in the traditional field, to take up cudgels against each other in preference to applying ten seconds' thought to the issue at hand. With the possible exception of the bloody field that is the Considerations before Judgement, nowhere is this more marked than in the question of whose location we use to cast the horary chart: the astrologer's, or the querent's. If astrologer and querent are both in the same place, as would most usually, but by no means exclusively, have been the case for Lilly, we have no problem. As today many consultations take place by phone, post or e-mail, it might seem that we have a valid point of debate; but if we adopt the traditional First Rule of Astrology – "Switch on brain before starting" – we find we no longer have a problem, no matter where our querent may be secreting himself.

The horary question is a request for information. It does not, in the terms of classical logic, exist until it meets the person who is able to provide this information. Until that point it is a no-thing, with no more real existence than any of the other ephemeral trash that floats through our heads. It is only when question meets astrologer that it becomes real.

This can happen only at the astrologer's own location, unless he is in the habit of leaving his brain at home when he goes for a walk.

We may compare the horary chart to a nativity. This is commonly cast for the moment of birth. There exist techniques for rectifying this to the moment of conception. No one, so far as we are aware, has seriously suggested casting the chart for the instant at which the particular spermatozoon that fertilised the egg was generated in the father's body; yet this is the equivalent of taking the querent's location as the basis for our horary chart.

So much for theory. We might also refer to practice. Our forebears did not have the luxury of computer atlases which can set a chart for any location at the touch of a button; nor even the capacious tables of houses with which we are familiar from *Raphael's Ephemeris.* A chart for a location other than his own would have involved the astrologer in a deal of work. For a nativity, this would have to be done. For a horary, he would undoubtedly have applied the Second Rule of Astrology: "Avoid unnecessary effort wherever possible". Our predecessors and models were professionals. The hallmark of a professional is efficiency of effort. We suggest that those who are so ardent in their claims that the querent's location should be used might spend a day or two calculating Regiomontanus cusps for distant locations from scratch. We should be more than surprised if this does not lead them to revise their opinions.

The Numbers Racket

Lilly gives us a list of various essential and accidental dignities and debilities, each numbered on a scale from 1 to 5, which scale originates with the Arab authorities. This has proved a snare in which those of Virgoan nature persistently entangle themselves, having in many cases been enticed inside by software programmers who know no better. This list is neither exclusive nor definitive, and should not be treated as if it were: its only purpose is as a rule of thumb guide to assist the student in appraising planetary strength. The only meaningful way in which such strength can be expressed is not '15' or '7', but strong/weak/middling, which, most particularly in horary, frequently reduces still further into the simple division of whether a planet is or is not *strong enough.*

We must first note that there are many points worthy of consideration which are not listed; to avoid being tedious, we shall mention just a few that come to mind. A planet gains strength through being in the house of its joy, and loses it through being in the house opposite. It gains by having north latitude and loses by south latitude (north latitude bringing it higher in the sky, assuming we are in the northern hemisphere). It gains by being in its hayz – some of the Arab authorities, indeed, regard this as a necessity if it is to act. Algol, Regulus and Spica are not the only fixed stars of consequence. Besiegement by Venus and Jupiter is as strengthening as besiegement by Mars and Saturn is debilitating.

The second point vital for correct understanding is that the numbers given were originally intended only to demonstrate the respective importance of the factors in question; thus combustion or conjunction with Spica (at 5 points) is more significant than dignity by face or by falling in the third house (just 1 point). What must never, ever, be forgotten is that like so much in the texts, there is always the rider "all things being equal". We see this most clearly in the lists of aphorisms, for a good number of which it would be easy to cite examples which are patently absurd, did we not remember that they are given "all things being equal"; so also here.

Combustion, for instance, scores -5. Yet there is a world of difference between a planet which is 2 degrees away from the Sun and getting closer and one which is 8 degrees away and separating. Being slow in motion scores -2; but again, what a difference between a planet which is in station and one which is only slightly below its average speed having recently turned direct: we cannot seriously think they are both debilitated to the same extent. Even when we have two stationary planets, we must remember that first station is far more debilitating than second, and weight our judgement accordingly.

Let us run through the list, marking some of the pitfalls. A planet in the first or tenth house is given +5; but, as Bonatus explains and practice confirms, a planet gains this strength only if within a degree or two of the cusp. Any further within the house and its influence is strongly diminished. This applies to all houses. This is most strongly marked when the planet is in the house, but in a different sign to that on the cusp: it is almost as if it were not really in that house at all, but there on sufferance, like a soldier billeted on a peasant household. On this point, we refer in passing to the 'five degree rule': that is, that a planet within five degrees of the next house cusp is regarded as being in that house. As Dr. Reason and Dr. Experience explain, this is only if planet and cusp are in the same sign; if they are not, it doesn't matter how close they might be, the planet will still be regarded as falling in the previous house.

The Moon gains strength when it is increasing in light. The mere fact that it is increasing is not, however, the point at issue: it is how much light it has that counts, not how much it is going to have. So the Moon in separating trine from the Sun is far stronger than it is at separating sextile ("all things being equal"). From separating trine, the Moon continues to increase in light as it approaches opposition to the Sun; but as soon as it is within about eighteen degrees of full it starts to lose strength rapidly, despite its gain in light, until when at the full – with maximum light – it is as weak as when new. Similarly with its decrease in light (scoring -2): once it has left the area of 'combustion' around full, the Moon still has plenty of light, and so power to act: far more than when at her last dying crescent before conjunction.

Partile aspects with benefics and malefics are listed and scored for good or bad. This does not mean that a close, but not partile, aspect has no effect. Suppose our main significator were at 15.59 Taurus. Saturn is at 16.00 Taurus. The aspect is not partile, but is plainly of great importance – all the more so as it is immediately to perfect. We might take this opportunity to clarify the meaning of *partile,* and the whole question of degrees. An aspect is partile if the planets are within the same degree. This is not the same as saying that they must be within one degree of each other, for which reason the planets in our example above are not in partile conjunction. A degree is, literally, a step: no matter how close to the edge of the step we might be standing, the point of importance is whether we are or are not on the same step as the other planet. Only if we are is the aspect partile. The degrees of the zodiac are a staircase, not a slope.

If Regulus, Spica and Algol are the only stars affecting a planet's strength, we might wonder how Aldebaran, Castor, Antares and Lucida Lancis give 'admirable Preferment, great Honours, &c' to an extent three times greater than any other testimony. The other major stars must be included in our assessment.

Now on to the subject of mutual reception. We have already dealt with reception at some length,[1] so here will deal only briefly with the assessment of the strength of planets in mutual reception. The score-sheet mentions only mutual reception by sign (+5) and by exaltation (+4). This does not mean that receptions by triplicity, term or face are of

[1] [Included here: *Are You Receiving Me?*]

no value: they are. Nor indeed does it mean we can ignore mutual reception by detriment or fall. If mutual reception is like friendship, such reception by debility is like getting into bad company: we cannot pretend that our association with those nasty boys at the end of the street is doing us any good. These rank as negatives as strongly as reception by sign or exaltation do as positives.

Mixed receptions are not excluded (e.g. Mars in the sign of Venus while Venus is in the exaltation of Mars). It must be remembered that the difference between the various levels of reception is one of quality, not merely of quantity. Planets in mutual reception by exaltation partake of the nature of exaltation; this is sensibly different from reception by sign, beyond the mere difference in strength. So also with the minor receptions.

Most importantly, we must consider the strength of the planets concerned. To say that two planets in mutual reception by sign gain by +5 if both planets are in detriment is simply not good enough. In a just society, a week in the pillory would be the only fit remedy for such nonsense. Planets in mutual reception are helping each other. A planet can help another only to the extent that it itself has strength. So mutual reception with a planet in its own sign is a lot more helpful than mutual reception with a planet in only its own face. Some authorities state that planets in their detriment or fall cannot be involved in mutual receptions at all: this is probably overstating the case – my friend may not be able to help me much, yet I still appreciate his sympathy – but it is a good deal more accurate than according great power to any mutual reception simply because it exists.

Just as I cannot help my friend if I am weak, neither can he help me if I am too weak. A planet that is badly debilitated cannot make use of the strength it is offered by another, much as a sick man may be too weak to hold down the food that would nourish him and speed his recovery. I am about to play football; my friend lends me his £200 boots; but if I am a hopeless player, it does me no good at all.

The key to the assessment of receptions is to remember always that we are concerned with qualities even more than quantities. So long as we understand how the various permutations of strength and weakness affect the quality of the reception, correct judgement of its strength will follow naturally. This is not reducible to a simple numerical formula.

We have explained often enough the reasons and shall not do so again, but two other common illusions on reception bear refuting yet once more: a planet without essential dignity of its own is peregrine, no matter how much mutual reception it might have; planets in mutual reception do not swap places.

Hush Yo' Mouth, Mr Lilly!

Those of a delicate constitution might like to make themselves a reassuring cup of tea before reading further, but Lilly is mistaken on the question of hayz. This is something which he only rarely uses; though he feels he ought to give it a mention, his memory has let him down. He says correctly that a diurnal planet is in its hayz if it is in a masculine sign, above the horizon, in a day chart. The Arab authorities, however, who do use hayz, make it clear that a nocturnal planet should be above – not below – the horizon in a night chart. It is possible that Lilly has misunderstood Dariot, who does give the Arab version, but (at least in the translation that Lilly used) in confusingly tangled form.

Another dignity which Lilly mentions by rote without giving any indication that he uses is that of 'increasing in number'. The usual formula is "if a planet is direct in motion and increasing in number." The common explanation for the term – that the planet is moving from one degree to another of higher number, as it does in direct motion – renders the phrase so plainly tautological that we may wonder why he keeps repeating it. *Increasing in number,* however, means something quite different, as it refers not to a planet's position relative to the sign through which it is passing, but to its position on the epicycle. The point in question is whether the planet is approaching or receding from the Earth, not its direction of motion against the zodiac.

As with hayz, some authorities – notably Abu Ma'shar – place considerable emphasis on increase or decrease in number. Remembering the Second Law of Astrology, however, it is a consideration which we can do without. While such an attitude may seem offhand, it must seem so only to those who do not work with astrology as a daily practice. Had we but world enough and time, we could tease every nuance of judgement from every chart; we have not. So numerous are the tools at our disposal that no astrologer, no matter how painstaking, has ever used them all; nor is this a failing: there is no virtue in making judgement more complicated than it need be. We are returned to the point repeated so often by Mr Lilly: the absolute necessity always in our judgements of combining 'discretion with art'.

Some Notes Concerning Light

Let us consider a few points on light, that we may clear up some common misconceptions.

Whenever the prime importance of light in the cosmology is mentioned, it seems sure that there will be someone who vociferously claims that this equates the absence of light in an object with the impossibility of seeing an object. If light were so important, the argument runs, planets below the earth would have no influence, one consequence of which would be that traditional astrology should not recognise the aspect of opposition. There is, of course, a world of difference between an object that has no light and an object that has light but for one reason or another cannot be seen. Both of them are invisible, but this does not mean that they are the same: cats have four legs; dogs have four legs; this does not mean that cats are dogs – the fault in the logic should be obvious. Nor, then, does the importance of light imply that planets have no influence when it is cloudy – or when we close our eyes! (No, really – it is no good sticking your head under the pillow throughout your Saturn return.)

That light plays so integral a part in the cosmology is apparent from the very first words of God at the beginning of *Genesis*. All that is created follows from there. This account indicates, however, that light, and also night and day, have an existence separate from that of the luminaries: light is created on the first day, the luminaries not until the fourth. This distinction has important practical relevance to our astrology. As the commentators make clear, the light of the first three days of creation is, as it were, 'essence of light'. This is then bodied forth in the luminaries, reflecting the increasing materialisation of the Creation. The Sun, then, even though it can be considered for practical purposes the prime source of light in the cosmos, has something of the same relation to light that a book does to knowledge: the book may radiate knowledge, but it does not contain it; the knowledge is preexistent to the book and manifests itself through it. The difference between the two natures of light, substantial and insubstantial, becomes clearer when we recall that the light that is physically visible is, for all its beauty, but the poor cousin of that light which is, as yet, not. (We have an inkling of this difference in the phrase that so-and-so 'lights up a room'; but this light is not, unless he happens to be an electrician, one by which we can read.)

The planet, as a created object, exists in time. Dante, whom we may regard as authoritative, and who, of all people, knows a thing or two about light, tells us: 'the divine light penetrates through the universe according to the fitness of its (i.e. the universe's) parts in a way that nothing can hinder it.' What we are doing when we assess the *essential* dignity of a planet is asking: 'To what extent does the body of this planet at this time manifest the essential nature of this planet?' That is, how well is Mars being Mars or Jupiter being Jupiter?

We know that the malefics become more malefic when they are essentially debilitated. This is equally true of the benefics: a debilitated Jupiter or Venus shows 'a taint of poison', often a very strong taint. The essential nature of Mars and Saturn is not malefic, but becomes so only when the particular 'part' of the universe that is Mars or Saturn is unable to receive and hence manifest that nature to the full. It is as if the planet's capacity to receive this essential light varies, and what it does not receive it cannot give out.

When we assess the *accidental* dignities of a planet we are asking, 'How effectively and in what manner can this planet at this time apply to this particular place whatever amount of essential dignity it happens to have?' Broadly speaking, the essential dignity answers the question 'Is he the good guy or the bad guy?' and the accidental dignities answer the questions 'Can he shoot straight?' and 'Has he got any bullets?' A planet with strong essential dignity and a mixture of accidental dignity and debility might show that the good guy can shoot straight and has plenty of bullets, but has misunderstood the situation and is firing in the wrong direction. It often helps to reduce the intricacies of the chart to simple terms!

The planets 'work', as we have seen, on the level of essence. That is, for instance, the planet Saturn shares its essential nature with undertakers, parsnips and Bayern Munich football club, so as Saturn moves they too will move. Nothing in this world, however, has its essence of one planet unmixed with any other: everything here is a mixture of all seven planets, one or two of which predominate. The exact measure of the mix determines the different ways in which parsnips and Bayern Munich mirror the movements of Saturn.

As Dante says, 'nothing can hinder' this essential light. But the light of any one planet does not exist alone: we have the accidental dignities, which show, as it were, what position that light holds relative to the other spheres of the cosmos: the sphere of the fixed stars, the spheres of the other planets and the sphere of the Earth (i.e. in which mundane house the planet falls).

Let us simplify the picture again. Let us suppose a strongly dignified Mars: my country has declared war for the best of all possible causes. 'Nothing can hinder' this light, so the strongly dignified Mars arrives on my doormat in the shape of an invitation to enlist. But there are accidental dignities to be considered. Mars is conjunct a malefic fixed star: as I pick up the envelope an invisible orchestra strikes up the funeral march and the scene cuts to little Tommy playing war on the porch and falling 'dead'. Conjunct a benefic but martial star, the orchestra plays Sousa and the scene cuts to little Tommy with the 'enemy' falling in swathes all around him.

But look – there is another letter on the mat, perfumed and in a woman's hand. It says, 'I love a man in uniform.' Venus trine Mars. It says, 'Don't go – I couldn't bear to lose you.' Venus opposition Mars. Both essential lights get their message across; in one case the messages harmonise, in the other they don't.

I pick up the letter. Even though the orchestra is belting out the funeral march, as soon as I read it I stiffen to attention with desire *pro patria mori.* Mars on tenth cusp. I pick up the letter, trying to focus my eyes upon it through the fog of my breakfast bottle of whisky, and then collapse unconscious on the floor. Mars in twelfth house. In all these cases the essential light arrives unhindered; what else is going on at that time, which includes all aspects, not only those that are commonly listed as accidental dignities or debilities, shows exactly how this essential light will make its presence felt in this world of generation and corruption. None of this involves the kind of light that we can put through a prism. We are talking of celestial things; we must expect them to behave in celestial ways: the stars do not abide by the laws of Earth.

The singular beauty of light – as St. Basil says, 'Light is that one of all created natures such that the thought of mortals cannot reach up to enjoy anything more pleasant' – rests in its being closer, in even its manifest form, to essence than anything else in the material world. Robert Grosseteste expands on this:

> 'Light is beautiful in itself, since "its nature is simple and in every way homogeneous": therefore it is united with itself to a very high degree, and most harmoniously proportioned to itself by its

> equality. Harmony in proportion is what beauty is: hence even without the shapes of bodies light is beautiful, by its own harmonious proportion, and is most pleasing to the sight. That is why gold, without any carved decoration, is beautiful: because of its sparkling shine. And the stars seem very beautiful to the sight, though they do not show us any elegance in the arrangement of their parts or the proportion of their shapes, simply because they shine with light. As Ambrose says: "The nature of light is such that there is all grace in its appearance: not in its size, or dimensions, or weight, as happens in other things. It is light that makes the other things of the world worthy of praise."[1]

How privileged we, to work with so fine a stuff.

As we must know our tools, the science of optics is worthy of our investigation. Traditional optics holds that a ray extends from the eye to the object that is seen. Modern science disagrees, but the truth of the traditional view can be verified by simple experiment (insofar as experiment can verify anything). Try drawing a sleeping cat, for instance, and notice for just how long it remains asleep. An aspect is, literally (*aspectus,* in Latin), a glance from one planet to another; the traditional optics is why we have the idea of planets projecting rays as they cast their aspects. It is, after all, the Sun and the Moon – the distributors, not the recipients of light – that rule the eyes.

That an aspect is a glance explains why a conjunction is not technically an aspect. In conjunction, the two planets become one, and you cannot glance at yourself. That an aspect is a glance also reveals that an 'inconjunct aspect' is a contradiction in terms. If the planets are inconjunct, they do not behold each other; if you cannot see something, you cannot glance at it. This is not to say, however, that an 'inconjunct aspect' (for want of a better term) is utterly without meaning; but its meaning lies exactly in the fact that it is *not* an aspect. A recent horary was posed by a woman asking about her prospects with a man whom she had never met, but with whom she had developed a friendship on the telephone. Their significators were in exact quincunx, making the very point that the querent and her friend did not, in all literalness, behold each other.

John Dee, in his *Propaedeumata Aphoristica,* gives a couple of aphorisms relevant to this theme:

'The rays of all stars are double, some sensible or luminous, others of more secret influence. The latter penetrate in an instant of time everything that is contained in the universe; the former can be prevented by some means from penetrating so far.'

'The insensible or intelligible rays of the planets are to the sensible rays as is the soul of something to its body.'

The Nodes

The debate about the exact effect of the Nodes has been continuing since long before Lilly's time. Al-Biruni tells us that: "It is related that the Babylonians held that the ascending (i.e. north) node increases the effects of both beneficent and maleficent planets, but it is not everyone who will accept these statements, for the analogy seems rather far-fetched." The dispute is this: is it North Node good and South Node bad; or is it North Node increases and South Node decreases? Much of the time, the visible effect will be more or less the same whichever of the two options is correct. Sometimes, however, the distinction can be of great significance.

That the debate exists is evidence of an unhealthy feeding of the empirical into the field of knowledge. If we are to deal with the empirical – as sometimes, perforce, it seems we must – we need to be exacting in the quality control of our data. The confusion seems to stem from too many people noting, for instance, Saturn falling on the North Node and finding that it works in favourable fashion. "Aha!" they think, "North Node good." But if they have failed to qualify the condition of Saturn at the time, their conclusion is baseless. It Saturn were in dignity when this observation was made – in one of its own signs, perhaps, or in Libra – the result could well have been favourable, and yet this would not necessarily imply that the North Node is in itself fortunate; it could equally well be increasing Saturn's dignified good nature and would also increase Saturn's malefic nature if Saturn happened to be in debility. With so many variables to consider, correct conclusions can be drawn only from first principles.

So what is happening here? The Nodes are the two points at which the Moon's path around the Earth crosses the ecliptic (the Sun's path around the Earth). As the Moon's path is traced on a plane set at an angle to the ecliptic, it varies in latitude: one half lies north and the other south of the Sun's path. So sometimes the Moon is above the Sun's path and sometimes below it. The point at which the Moon's path crosses the ecliptic heading north is the North Node; the point of intersection when it is heading south is the South Node.

Change in latitude is significant: increase in north latitude is an accidental dignity. It increases the planet's power. South latitude correspondingly diminishes it. The more north latitude a planet has, the higher in the sky it climbs. The effect of latitude on physical appearance is that north gives fat, south gives lean. Whether the gift of fat is benefic or malefic will depend on whether we are discussing a pig or a supermodel.

Again we return to the distinction between essential and accidental dignity. The more essential dignity a planet has, the better it behaves: even a malefic begins to show its positive side. Accidental dignity gives a planet increased power to act. It puts it behind the wheel, but does not teach it to drive. As the Nodes offer accidental dignity and debility, and the nature of this particular dignity, as shown by its relation to latitude, is one of increase of power, we must conclude that what is related of the Babylonians is correct: North Node increases the planet's ability to act, whether for good or ill; South Node diminishes it.

While on the Nodes, we must also correct the common illusion about their aspects, and that pernicious idea of 'degrees of fatality'. The Nodes are not bodies: they are points in space. They have no light, and no existence in themselves. How can they possibly cast aspects? They affect a planet that is on them, but nowhere else. We might liken the North Node to a chair: a planet that stands on it is better placed for banging people on the head or showering them with largesse. The fact that there is a chair on the other side of the room does not increase my height unless I choose to go and stand on it. It cannot raise me up by aspect. Similarly with the South Node, which might be likened to a hole in the ground: I will not fall into it if I am on the other side of the field.

As for 'degrees of fatality': the point exactly midway between the Nodes is often of significance, but this is not because this point *does* anything at all. It merely marks a place, which is often found occupied, in horary, by the significator of someone at a major turning-point. We see no reason for regarding the degree of that same number in any other sign as having any special significance.

Give Me the Moonlight...

Why, we might wonder, does the Full Moon have no power? It has been increasing in strength as it increases in light, ever since it left conjunction with the Sun, yet as soon as it finally attains the goal it has been seeking, with its maximum amount of light, rather than finding its maximum strength it has none at all.

It is an awesome but uncomfortable truth that the cosmos is arranged according to the principles not of human sentimentality but of Divine justice. In a fragment from a lost tragedy, Euripides compares justice to 'Phoebe gazing across the heavens at her brother from the rosy flush of the clear morning sky'. That is, the Sun is rising and the Full Moon just setting. 'Neither,' he says, 'Hesperus nor Lucifer is so wondrous.' (Hesperus and Lucifer being Venus as evening and morning star respectively.) Now, Venus is a very fine thing: it is not without significance that it is the only star that casts a shadow, the only star by whose light we may read. Just as the Moon's light helps us manage in the absence of the Sun, so the light of Venus enables us to scrape along in the absence of both Sun and Moon. Beautiful though this planet is, beautiful as are the promptings of 'rightly ordered love', however, it is not the match of justice as signified by the Full Moon, reflecting to its maximum capacity the light of the Sun.

Even from the most popular of the modern texts, we are familiar with the idea of ever-hungry Cancer; ever-hungry too is its ruler, the Moon. As is immediately apparent from a glance into the night sky, it is always, except at full, lacking something or everything; it desires what it has not. We might say that the Moon is filled with greed. The waxing Moon can be characterised as greed for what it has not yet got; the waning Moon as greed for what it once had but has lost. At Full, it is full: its greed is momently satisfied. It has all that it can want, its capacity is filled. With its greed at last sated it has no power, for its whole motive force has gone. 'It is greed alone,' Dante says, 'That perverts judgment and obstructs justice'; so only once this greed is filled does the (now-Full) Moon become this symbol of justice. It is at Full that the Moon no longer has desires of its own, but can become 'the handmaid of the Lord', as is reflected in the pattern of planetary joys.

We see here also how Mars can be at once dignified (triplicity) and debilitated (fall) in Cancer. For this endless greed is productive of desire (Mars), and yet is at once obstructive of it as well: what wants everything does nothing. The Moon makes a poor job of organising Mars, as if he were a great warrior reduced to selling his sword in ignoble causes.

Timing and Precision

If we are making predictions from the horary chart, how precise do they need to be? The chart does provide us with a great deal of rope, so here in the workshop much effort has to expended in order that the younger apprentices may be prevented from hanging themselves with it. In the intoxication that comes with the dawning realisation of how potent a tool horary is, there is always the temptation to write a novel when we should be sending a telegram. If the question is "Does he love me?" the querent wants to know whether or not he loves her; that the astrologer might feel able to discourse on the competence of her plumber from information given in the same chart is doubtless very impressive, but neither of interest nor necessary. Horary is, as we have remarked so often before, like surgery: if we are operating on the heart we do not need to whip out the appendix just because we happen to have the body open.

If we have to break the news that "No, he can't stand you," it is not unreasonable that we should have a quick look round the chart to see if we can find a glimmer of hope for the future: having our clients queuing at the Tallahatchee bridge is bad for business. That said, the general principle of practical horary – horary as practised, rather than horary as theorised about by people who don't do it very often – is 'answer the question and then stop'. This is not simply a matter of avoiding unnecessary effort; but just as the more of his bodily parts we snip off the less likely our patient is to survive, so the farther we drift from the immediate point of the question, the more likely we are to go wrong.

The reason for this is that we have but twelve houses to describe everything in the universe. The house meanings that are implicit in the question are clear: the querent is the first, the person who may or may not love her is seventh, for instance. Major aspects or receptions between our main significators and the rulers of other houses are usually, but by no means always, clear enough in their meaning. But let us begin to paint in every tiny detail of the situation, and each house, and therefore every planet, has an infinitude of possible meanings, revealed as we turn and re-turn the chart. In plumping for any one of these rather than any other we are doing exactly what we most wish to avoid: turning the chart from a mirror of truth into a mirror of the astrologer.

As Sam Goldwyn said, "If you want to send a message, use Western

Union." We do indeed want the message. MGM we are not; Western Union we are.

So we should stick to the point. But how much precision do we need in what we say about that point? Lilly, being a true Taurus, reminds us repeatedly of what is 'too scrupulous a Quere' or 'too nice (i.e. fussy) a poynt in Art'. The purpose of the exercise is to give a satisfactory answer, not to demonstrate how clever we are: let us leave that to David Copperfield.

As Lilly was well aware, there is a sound practical reason for this 'big picture' approach. If I predict that you will marry in the summer of 2030 and you do, you think I am a great astrologer. If I predict you will marry on July 5th, 2030 and you marry on July 4th, you think I am wrong. If the prediction is for next week, it is not unreasonable to expect that we give the exact day. If the prediction is for ten years' time, to attempt this – certainly from a horary – is to overreach ourselves, and overreach ourselves for no purpose. Similarly with questions on the exact number of children, or the exact amount of money: 'lots' or 'more than you expect' or 'next to nothing' are quite satisfactory answers. That we do not attempt more is not failure, but knowing our just limits.

This is not to say, however, that horary is not able to provide a remarkable degree of precision, as this example demonstrates.

The Apprentice was waiting for a new fridge to be delivered and cast this chart for the question, "At what time this afternoon will the fridge arrive?" He expected to see a planet applying either to the Ascendant or to Jupiter, ruler of the Ascendant, both of which signify the querent. The chart thinks differently, however, and shows the event elsewhere.

A planet on an angle is one of the ways a chart has of shouting "This is important – look over here!" so the Moon must be considered. The Moon would anyway be a good planet to take as ruler of refrigerators ('white goods'), and it is also relevant as ruler of the sixth, the house of 'slaves and servants': the fridge is not going to deliver itself. Whichever way we look at it, the Moon is significant.

It applies to the seventh cusp, rather than to the first, which seems to reflect the situation: try as the Apprentice might, he could not summon the same degree of interest in the fridge that Mrs. Apprentice displayed (she will not, he tells me, even allow him a corner of her new acquisition in which to store lunar aspects, despite the speed with which they deteriorate in hot weather). So even though the chart does not behave as expected, it still clearly shows the event.

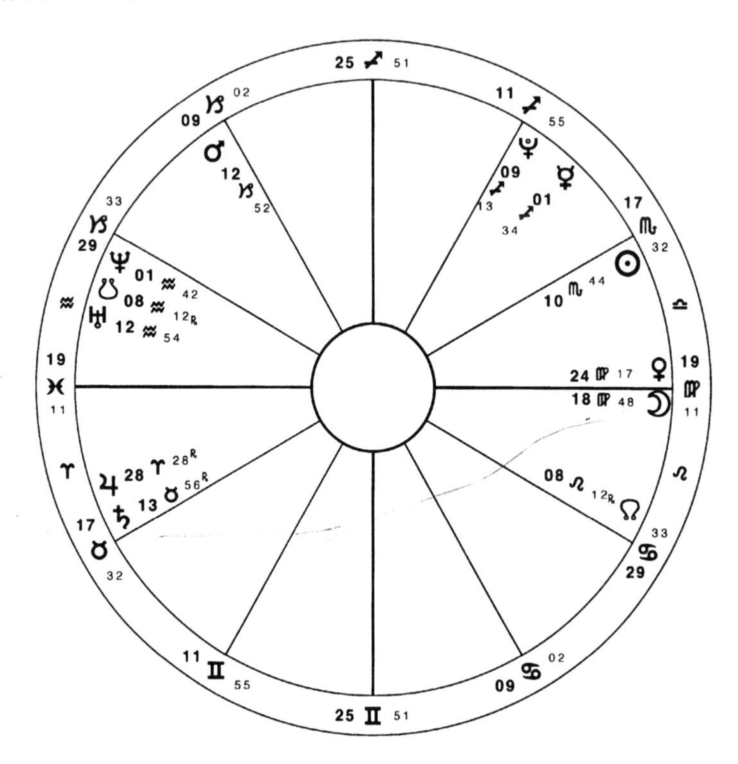

Chart 7. When will the fridge arrive? November 3rd 1999.
2.52 pm GMT. London.

The Moon is almost exactly on the angle – just twenty-three minutes of arc away from it. Looking for seconds in the chart is foolishness; twenty-three hours was out of the question, as in that case the chart would have shown non-arrival that day. So these twenty-three minutes of arc must show twenty-three minutes of time. The chart was cast at eight minutes to three; the doorbell rang at exactly fifteen minutes past. Perfectly exact and beautifully simple.

The Seasons

We have been asked to shed some light on the puzzle of how our antipodean chums manage to have summer while the Sun is in Aquarius. The explanation hinges on the difference between true spring and local spring. To those still persisting with the modern imitation of thought this makes little sense, but once the distinction between essence and accident has been grasped it is perfectly straightforward.

Robert Grosseteste answers this question in his *Hexaemeron.* This was written while the antipodes remained a theoretical possibility rather than a material discovery, but the principle is unaltered.

> 'These periods (i.e. the seasons, as marked by the Sun's movement through the zodiac) marked out in this way by the movement of the Sun, are the four seasons of the year, for the whole world without qualification. But with regard to the individual parts of the inhabited world spring is said to be that quarter of the year that is more temperate with regard to heat and moisture.... these parts of the year, considered according to these qualities and effects, begin and end at different times in different places... But in relation to the heaven and to the earth without qualification, as we have said, these four seasons of the year always begin at those four moments when the sun begins its journey from the four points of heaven, i.e. the two equinoctial points and the two solstice points.'

That is, what we might loosely term 'essential' summer begins when the Sun enters Cancer, even if you are knee-deep in snow at the time; while Kylie and Bruce are eating Christmas pudding on the beach in the middle of 'essential' winter, even though locally it is hot and sunny.

The distinction between essence and accident, to which we keep returning by different paths, is of crucial importance. Much of the failure of modern astrology can be directly ascribed to its attempt to root itself in the world of modern 'science', a world which pretends that essence no longer exists.

Aldebaran

In the traditional imagery, the bull of Taurus presents his left eye to us, this left eye being the first magnitude star Aldebaran. The bull, in its fixity and earthiness, is representative of matter. Its left eye would be ruled by the Moon (Moon rules the left eye of male creatures) and as such shows the soul entering Creation, which is why Aldebaran is so very important a star. The right eye, which we cannot see, is looking away from us, out from the spheres of the cosmos towards the Divine. This is the Sun – the Spirit – ruler of the right eye in males, while the Moon is focussed on this world of generation and corruption. It is, Ficino explains, only when the soul turns away from the light by which it perceives things of a higher order (Sun) and concentrates its attention solely on the light through which it perceives things of the lower orders (Moon), that it falls into the body.

The bull-leaping of ancient Crete mirrored the soul's passage through incarnation. In the pictures that survive, the figure facing the bull is the soul about to become incarnate. The figure above the bull – typically shown upside-down and having assumed the colour of the bull – is the soul in the midst of life. The figure behind the bull is the soul having negotiated life. Not everyone, of course, gets their soul out of life in the desired condition. The Athenians were required to send seven youths and seven maidens to Crete. Seven, being the number of manifest creation – hence seven planets – and given here in both genders, shows that it is *everyone* that is required to go. They are pushed into the labyrinth where, lost, they are devoured by the bull: the soul overwhelmed by life. The hero overcomes the bull of matter and returns home as king. Much the same is seen in the ideal behind the *corrida,* where the hero, in his suit of lights (the multi-coloured raiment of the planets) so masters the bull that it has no recourse other than to cry out 'matame' – 'kill me'.

Uranus

Pointing out errors in the thinking behind modern astrology is like shooting fish in a barrel; the sense of sport quickly fades. But sometimes we come across a point that has a wider relevance. Such a one is the attribution of myth to the supposed meaning of the planet Uranus. Indeed, the use of myth to comprehend the significance of any planet is to be undertaken only with extreme caution: it is necessary that we raise ourselves to the level of the myth rather than following common practice by reducing the myth to fit our own small thoughts.

With other planets we can quibble over specific references to myth. The application of myth to the understanding of Uranus, however, shows a deeper misunderstanding, not only of the myth but also of the fundamentals of astrology.

The myth tells that Uranus was castrated by Saturn. This relates to the distinction between the plenitude of potential, as indicated by the signs of the zodiac, which contain all possibilities, and the limitation of the actual, as indicated by the planets. To continue the procreative theme, it is the distinction between the millions of spermatozoa, all bearing the potential for life, and the single spermatozoon that brings its potential into manifestation.[2] The reduction of all possibility to just one actuality is the castration – hence the restrictive nature of Saturn with which we are familiar.

The zodiacal signs contain all possibility; from this, the planets spin the slender thread of what happens. The point of this myth is precisely that Uranus is not a planet, while Saturn is; to identify the Uranus of myth with a planet is quite erroneous.

The distinction between potential and actual is a fundamental frustration in the human make-up: the ubiquitous desire to have the cake and eat it too – to have both the potential and every specific actuality, whether this is manifested as the truck-driver's lament 'so many women, so little time', or the great-souled Alexander weeping that there are no more worlds to conquer. To distort a myth of such profundity in order that we may propagate our preconceptions through some unsuspecting planet does us no service.

Fixity and Angularity

As various points of technique show us, the fixed signs equate with the angular houses. In the timing of horary predictions, angular houses and fixed signs both show a slow time-unit. In questions of truth, angular houses and fixed signs both carry the sense of solidity that we want in order to judge something true. This equation is puzzling, as the angular houses 'ought' to equate with the cardinal signs.

Our understanding is not helped by the modern chart-wheel. This naturally inclines us to see the houses in a progression from first to twelfth, with the result that we see the cadent houses as tagging along behind the succedents rather than in their true position as falling away from the angles. In a structural sense, as Morinus emphasises, the chart is built from four groups of three houses, each group being centred on an angle. The succedent house follows the angle and the cadent house falls away from it. The traditional square chart makes this much clearer.

That this is the structural reality of the chart does not in any way negate the truth of the pattern, particularly important in medical astrology, of signs through houses which does locate the cardinal signs in the angular houses. The sense of solidity that gives us slow timing and reliability in statements is as real as is the same sense of solidity in the

[2] We live out this myth week by week as from all the millions of lottery tickets, each carrying its dream of new life, only one brings this dream into reality.

main beams of a roof, and in our chart this is located in the angular houses and the fixed signs.

Although the traditional manner of calling both signs and houses 'houses' can lead to confusion, we lose much by our exaggerated distinction between the two. It is well to remember that they are indeed both 'houses': celestial houses and mundane houses. It is well also to remember that our perception of reality is based in the mundane houses. As such, it is at best only partially true.

We might consider a king, who points to one of his courtiers and says: 'Go ye and discover new lands.' This is the initiating impulse, corresponding to the cardinal modality. The exploratory impulse, corresponding to the fixed modality, is when said courtier leaps ashore and claims this land in the name of the king. The mutable, double-bodied, modality, combining the results of the cardinal and the fixed, is when the courtier returns home laden with gold, silver and potatoes. This is the 'true story': the story from the perspective of the celestial houses (the signs).

From our fallen, mundane perspective, however, things are different. From our point of view, the initiating – cardinal – impulse is when a strangely attired chappie turns up on the beach, shouting unintelligibly and acting as if he owns the place. This is the start of the cycle in the mundane sphere: the first house. We see nothing of the true initiating impulse. The mundane cardinal, then, (first house) equates with the celestial fixed.

Where's the Air in Aquarius?

A source of confusion for all beginners in astrology, and often a puzzle for those who have learned more, is the mystery of why the only one of the twelve signs with any visible water in it is not a water sign. The distinction is in the quality of the water: this is sweet, while that inhabited by crabs and fishes is salt.

The importance is emphasised by the familiar image of the water-bearer. These images are not pieces of random creativity, but are carefully constructed symbols which repay study. The water-bearer is not merely carrying, but pouring out, his water from a pitcher that is balanced on his shoulder and is thus on the same level as his head. The water flowing from this level indicates the sweetness of reason – something far different from the passional medium in which crabs and fishes dwell, and befitting Aquarius' nature as the most humane of signs.

A similar example of the language of symbolism is found in the Elizabethan dances with which films will have familiarised our reader. The partners hold hands at head height, symbolising the rational (in the finest sense of the word, as that which distinguishes man from beast) nature of their attachment.

It Don't Add Up!

OK: you can add up. This is very clever; but you do not need to demonstrate this talent on every possible occasion. So stop it!

There is a veritable obsession with adding things up in the world of modern traditional astrology, fuelled by the software writers who lose no opportunity for showing off their arithmetical prowess whenever confronted with more than one number on a piece of paper or computer screen. The question of whether there is any meaning behind this barrage of arithmetic seems never to have occurred to anyone. We even see such statements as 'its strength is 6' or 'Jupiter is 5': statements which mean absolutely nothing.

Mr. Lilly has a lot to answer for, putting those numbers in his lists of dignities and debilities! But, it should be noted, nowhere does he describe the condition of a planet by saying 'its strength is 7'. These numbers are a rough guide to the relative importance of the various dignities and debilities. That and that only. Yes, Lilly does add these numbers when discussing nativities; but we must remember that Lilly stands at the juncture between the real world and the modern world. On the plus side, this having one foot in our world is why he speaks so clearly to us; on the minus side, it does make him prone to many of our modern errors. Lilly, for all his greatness, is rather less traditional than we think.

These numbers cannot be meaningfully added together for the following reasons:

1. things can be added only if they are the same. 2 cats + 2 cats = 4 cats. But 2 cats + 2 tables = 2 cats + 2 tables (and 2 dogs + 2 sausages = 2 dogs). In the world in which we live, besotted as it is with quantity, it is forgotten that we cannot add things that are different in their quality. The various dignities and debilities differ in their quality; that is why they are various. They are not only stronger or weaker than each other, but sensibly different in their effects.

It is true to say that, whatever their qualitative differences, the dignities all strengthen to some extent and the debilities all weaken. But this still does not mean that they can be added (i.e. in the same way that we can add cats and dogs, if we call them both 'animals'). If, for instance, malaria is 5 of an affliction, and a cold is 2 of an affliction, having a cold and malaria at the same time does not render me 7 afflicted. Or if braces keep my trousers up 3 strong and a belt keeps them up 3 strong, having both does not mean they are kept up any better than having just one.

If, for instance, I am cazimi I can do what I want. Being swift in motion, sextile Venus or increasing in north latitude does not affect this one jot. It will alter the way in which I express this ability to act, but does not give me any greater ability. So also if I am combust: this is as bad as it gets – being slow in motion is neither here nor there. It tells me I am likely to be combust for longer, but does not weaken me any further. And so with whatever other examples we may choose.

2. Most of the various terms are themselves variable – yet our maths fiends ignore this in their dash to the abacus. Lilly rates combustion as -5, meaning that it is a major affliction. But there is a world of difference between a planet that is two degrees from the Sun and getting closer and one that is seven degrees from the Sun and separating. To class them both as '-5' contradicts any claim we might make to subtlety of judgment.

3. Lumping together essential and accidental dignities is only a special case of point 1, but a particularly parlous one. Anyone caught doing it in the workshop gets to clean out the cattle stalls for the next month: an object lesson in accidental debility if ever there were one!

When handling the dignities and debilities we have two options. We can carefully analyse the effects of each individual dignity when assessing how that planet works. Most of the time, this is not necessary: we pick out one or two qualities when considering the planet and ignore the rest (just as if I want my friend's opinion on something I will ignore the fact that he has a broken leg as it is not pertinent; if I want him to run an errand for me, it is very pertinent indeed). Or we can give a broad assessment of the planet's condition. This should rely on the traditional technical terms 'good', 'middling' and 'yuk'. If we think we achieve greater precision than this by arithmetic, we are mistaken.

"*Most horary texts indicate that if you ask a trivial question you'll get a trivial answer, or unless you have a burning question the answer you get may be inappropriate or unhelpful somehow.*"

In this way horaries are like dreams – in the way that it is immediately sensible how some dreams are of great significance. There are horary charts like that, and they tend to be the ones where the querent has a pressing and serious need that has almost crystallised their life around itself. But this certainly does not mean that only 'important' issues can be handled.

Lilly himself says that when he was at a loose end he would ask someone to hide something and then 'in merriment' set a chart to find it. It is important to put this in perspective: in one way, viewed from the level of the celestial spheres even our most significant concerns are of the utmost triviality. But seen in that way, there is so little distinction between the rise and fall of empires and what time the plumber will come, they both being quite insignificant, that we can as well answer one as the other. On the other hand, we are told that every raindrop has an angel assigned to guide it to its destined place of rest, so even our most trivial concerns are of absolute importance.

And who is to say what is trivial? When one of the workshop cats goes missing this is pretty important! Far more so than many an event which arrives with its Certificate of Importance duly stamped, and yet which is far lower in any reasonable scale of priorities.

"What are we to make of the fact that Lilly's judgements are sometimes based on aspects that more accurate calculations show are separating, when he says they are applying?"

In several of Lilly's example charts judgement rests on an aspect which he took as applying, but which modern computer calculations show as being already separating. The first, more minor, point is that our absolute faith in the accuracy of computer calculation must be questioned. We see an extreme example of this in the dismissal of certain observations in the most ancient records which do not fit the current view of where certain planets ought to have been at certain times. When – as most of the time they do – the observations recorded fit the modern model the ancient astronomers are clever little chaps, doughtily making accurate scientific records despite living in an age of darkness and superstition. When the observations do not fit the modern model, however, our predecessors are relegated from 'doughty little chaps' to 'purblind amateurs, who can't tell one side of the sky from the other'. If ancient perception differs from modern theory, it is not necessarily ancient perception that is wrong.

Lilly did, however, show an admirable disregard for the niceties of calculation. Consider his practice: this was not the world of the modern astrologer, where clients are infrequent and each gets a brand new chart at the click of a mouse. Lilly would enter his consulting-room, where there would be a queue of people waiting to speak to him. If he were feeling especially virtuous – which, with the amount of practical Taurean influence in his chart was probably not that often – he would set a chart before he began work. More often, he would look at the chart he had used the day before and quickly make a few adjustments. These adjustments would more usually be off the top of his head than through careful calculation: 'Mercury was at 7.20 Virgo yesterday morning – let's knock it forward about a degree and a half. 8.50 Virgo: that's close enough.' Having set his chart for the day he would then push this round by whatever he considered a suitable amount for each new client. He did not have a 'Now' button wired through the internet to an atomic clock. We do not know what form of time-keeping he did use, but it is most unlikely that his choice of minute was anything other than an approximation: 'It was 11 o'clock a little while before the last querent came in. She rabbited on a bit. I'm getting hungry. It must be about 11.35.'

This does not, it must be stressed, mean that his charts were any less valid than either the charts we cast today or our attempts to reconstruct the charts that Lilly cast. We must always remember that the astrology and the astrologer are not outside the life. Tempting though this illusion might be, it is still an illusion. For a horary question the querent, whether through his own will or through an apparently random event, such as the arrival of his letter at the astrologer's lair, selects a time for the question to be asked. He also selects an astrologer, and he selects that astrologer at that time (e.g. if the astrologer is having a bad day, as the astrologer is fully entitled to do, the querent has chosen that astrologer when he is below his best). Querent, astrologer, question and chart are all part of the same reality, so the chart – even the 'wrong' chart – is intrinsic to that moment, and hence to the question asked at that moment. So Lilly was perfectly correct with his 'inaccurate' chart, just as, if it were somehow possible for a modern with a computer to be asked a question now at the same time as one of Lilly's charts, his, different, chart for the same moment would also be perfectly correct.

"I understand the condition of cazimi (to sit with the king) but why are other conjunctions and oppositions to the Sun harmful?"

Getting too close to the king is a risky business: you are safe if you are in his very bosom, but live on the edge otherwise. Being opposed to the king – or to the 'Lord of Life' that is the Sun – is bad news. Note, for instance, that Mars, Jupiter and Saturn are retrograde when opposed to the Sun, i.e. opposition to the Sun automatically puts them into a bad state, without it even being remotely close.

That fine but neglected teacher of astrology, Mr. Shakespeare, explains the distinction between combustion and cazimi most admirably through the mouth of Portia:

> *His sceptre shows the force of temporal power,*
> *The attribute to awe and majesty,*
> *Wherein doth sit the dread and fear of kings;*
> *But mercy is above this sceptred sway,*
> *It is enthroned in the hearts of kings,*
> *It is an attribute to God himself...*

So what we get in combustion is the fearful sceptre; when cazimi, in the heart of the king, we find mercy.

3

The Master Astrologer: William Lilly

ASTROLOGY ON THE BARRICADES

Above the bench where the Apprentice sits whilst polishing his master's astrolabe hangs a signed photo of one of the great English astrologers: Nicholas Culpeper. Although he is remembered mainly for his *Herbal*, his outspoken opinions and bizarre sense of humour make his astrological writing as entertaining as it is instructive.

That the *Herbal* seems today the most innocuous of books, a suitable gift for any maiden aunt, makes it hard to realise that its publication was a revolutionary act. This was, remember, in the days when refusing to tip one's hat bordered on the treasonable and politics had yet to become something indulged in only by foreigners. Culpeper's stated aim was to enable every man to be his own physician by providing him, at a very reasonable price, with all the requisite knowledge. That would have been crime enough; but this knowledge was not obscured by being written in Latin, being, for the first time, set forth in the common tongue.

His work to this end, of which the *Herbal* was a part, was damned, attacked not only by the College of Physicians, whose excessive earnings he threatened, but by the whole battery of the religious and political establishment. The public was warned that every one of his prescriptions was mixed 'with some scruples, at least, of rebellion and atheisme'. In the Seventeenth Century, science was not as free of political and social constraint as it is today.

Culpeper's astrological textbook is his *Decumbiture*, a decumbiture being the time at which a sick person takes to his bed. It was common practice for the physician to cast a horoscope for either the moment of decumbiture or the moment at which he received a sample of the patient's urine. From the chart and consideration of the urine and any other bodily effusions – into which Culpeper delves with the greatest delight – he would diagnose, prescribe and prognosticate. This method had the virtue, now again becoming more popular, of treating the patient rather than the illness – of regarding, that is, the patient as an individual organism with individual qualities, not as one of a race of identical machines which has developed a mechanical fault.

Although the book is concerned primarily with medical astrology, there is much of general application. He valuably points out, for example, the extreme debility of the Moon in Gemini. Watching the progress of his patients' illnesses provided ample opportunity for refining his theoretical knowledge in the fire of practice. Nor must the entertainment value of the book be discounted: it is the forerunner of every medical black comedy you may ever have seen.

Lilly the Pink?

Autobiographies must always be read with caution. This is all the more true of William Lilly's: the autobiography of an outspoken supporter of a failed revolution, written under the restored crown. Yet almost all we know of Lilly's life comes from his autobiography.

When writing his life story, after the Restoration, Lilly played down his efforts in the Parliamentary cause. His apparent apostasy has been criticised by historians; but if the greatest threat to one's safety is through one's tea arriving a few minutes late, it is easy to criticise. Lilly escaped execution only through the intervention of well-placed friends: many whose actions had been far less suffered far more.

So how far left was William Lilly? Was he really in the pay of the Kremlin, or just an imperialist stooge? Evidence is nothing but circumstantial; but, speculative though it is, we can draw some reasonably convincing conclusions.

It is the fashion among historians to assume that, as an astrologer, Lilly was necessarily of dubious moral character, and therefore his support for the Parliamentarians was that of a man following the main chance. This is confirmed by the way he plays down this support in his autobiography. Again, this judgement is easy to make from the safety of a university cloister. That Lilly had an understandable desire to keep his head in conjunction with his shoulders, and was prepared to moderate his public pronouncements in order to increase the likelihood of this happening, does not prove he was a fraud. Lilly was living in the real world, far from the tranquil groves of academe.

It is easy, but badly misleading, to view the Civil War and the Interregnum from a modern perspective as integral, unchanging events. From the very start, everyone is assumed to have known just what was going on, just what they thought about it and just where it all would lead. But those who lived through these times lacked the clarifying benefit of hindsight. Their views would necessarily change and develop as events unfolded, making some positions untenable and revealing new possibilities. Even astrologers can work only within the limits of apparent possibility: in 1642, when the war started, the execution of the king was undreamed of, so it is unreasonable to expect any astrologer to have seen it in a chart. Lilly's views, like those of anybody else, changed, and not only with developments in the political situation, but also with the passage of time through his own life. These events took almost twenty years; he was not the same man at the Restoration as he was when King Charles raised his standard to begin the war.

There are few historians who avoid this trap of assuming that their subjects held consistent ideas throughout this twenty-year period. Christopher Hill points out that it is – which should be no surprise – possible to consider at least some forms of the radicalism of that age as the revolt of teenage boys against their parents. Had Charles I had the wisdom to provide each of his opponents with a Nintendo and a new pair of Reeboks, the course of history could well have been different.

In this context of teenage rebellion, it is, as an aside, worth noting the copious consumption of tobacco by many religious radicals of the time: tobacco, still a novelty drug, being regarded as a means of increasing spiritual awareness, and religious and political radicalism going hand in hand. One vicar even had a reputation for smoking his bell-ropes when he ran out of tobacco.

We should not, therefore, unlike some of the commentators, be surprised if Lilly's astrological judgements on the Civil War and the fate of King Charles did not leap fully formed and perfect from his head at the moment when war began. The astrologer relates symbols to reality; he does not relate them to fantasy. Only when certain hypotheses began to coalesce into real possibilities could he include them in his judgements. It is for this reason, and an understandable caution regarding his own safety, that his pronouncements became stronger and more overtly anti-royalist as the war progressed, not through a following of fortune.

The evidence

We have political histories, social histories, economic, intellectual and military histories; if ever someone should write a spiritual history, the Civil War will be seen as a turning point, forming the mould whose shape English society has borne ever since, and with no small effect on the world beyond these shores. We can begin to piece together Lilly's part in this.

One reading of the war sees it in the context of an economic struggle between the more modernised – I hesitate to use the word 'advanced' – south and south-east, and the comparatively old-fashioned north and west. Even without the evidence of his writings, we could be fairly confident that someone of Lilly's particular class, living in London, would probably have stood, in anachronistic terms, somewhere left of centre. His religious faith was clearly millenarian, though its exact form is unclear, as he looked forward eagerly to the day when kings would become as mere men. He would have had far less reason to exaggerate his radicalism during the ascendancy of Parliament than to conceal it during the Restoration, so the general impression his writings give of an eagerness to see the last king strangled with the guts of the last priest probably reflects his true beliefs reasonably accurately. His supposed sympathy for King Charles as an individual human being, while apparent, seems to have been over-emphasised. He might not have been willing to wield the axe himself, but was surely not sorry to see it fall.

The execution of the King was far from being a purely mundane event, like the removal from office, or even the assassination, of a modern head of state. His death was widely seen as a necessary step towards the creation of the Kingdom of God on Earth, the Second Coming of Christ, or some such – the exact emphasis differed from sect to sect. It is the error of a secular age to see the Civil War in purely secular terms, a necessary step towards the creation of a constitutional monarchy. In so far as a division can be made between them, Lilly's attitude to the execution is a product more of his religious than his political beliefs.

Like Culpeper, Lilly was in the van of those ushering knowledge from the shadows of dead languages into the light of the vernacular, though the lack of an astrological establishment in the way that there existed a medical establishment enabled him to avoid the opprobrium heaped on his contemporary; Lilly was threatening no one's earnings so much as his own. Not only the language in which he chose to write, but the attitudes implicit in his writing betray his inherent radicalism. Not for him the elitist role of omniscient master of the arcane, deigning to disclose his secrets. He describes himself as 'student', not master, of astrology, and after weighing up various differing theories is not above shrugging his shoulders and admitting 'I don't know' – even about such important matters as the nature of Fortuna.

Interesting, too, is his growing retreat into the practice of medicine. Even in *Christian Astrology*, a work from the spring of his career, he devotes more space to medical than to any other horary questions. After the Restoration, more and more of his work was as physician rather than astrologer, treating the poor – as did Culpeper – for little or no fee. This may not seem a radical action; yet consider how hot a political potato is free medical treatment, even today.

We may wonder if Lilly turned from astrology in disillusion; if looking into the future had lost all point after the refusal of the Kingdom of God to manifest in England. The failure of the revolution, its collapse into greed and property squabbles, the suppression of the radicals, must have been shattering. He was far from the only, and far from the last, disappointed revolutionary to turn from curing the body politic to curing the body carnal.

Most curious of all is his removal from London to Walton. Lilly did not make the final move until 1652, but had spent much of his time there since 1638 or before. Nowadays, Walton is cosily conservative; yet, remarkable as it may seem to any who have attended the annual pilgrimage of Lilly Day, it was once the most fervent hotbed of radical politics in England.

While Lilly was writing *Christian Astrology*, also resident in Walton was Gerrard Winstanley, the leading light among the Diggers – in modern terms, for want of a closer approximation, anarchists. The first practical Digger community was established just down the road at St. George's Hill. Winstanley had high regard for astrology, it being one of the few subjects which were to be taught in his ideal community. It is hardly likely that he would not have been client to so eminent an astrologer sharing so small a home town. This is no reason, of course, to assume that Lilly agreed with his views; he may have detested the man and all for which he stood. He may have moved to Walton for a thousand reasons unconnected with politics. But this juxtaposition of two figures, so much of whose lives remain unknown to us, is intriguing. Were the Digger experiments set up with advice from Lilly? There is nothing in his work to deny the possibility – though, this being pure speculation, nothing either to confirm it.

Of one thing we can be certain, however: the great disservice that Lilly did to astrology. Unbeknown to him, his fervent support for Parliament was one more factor helping to dissolve the bond between human and divine, without which astrology loses all credibility. The execution of the king, which fatal blow killed also the sense of divine involvement in the mundane affairs of man, and hence isolated the spiritual in a sphere apparently removed from daily life, was a milestone on the path to the materialist society we enjoy today. In this society the underlying assumptions of astrology, which can be neither weighed nor measured, cannot be understood: here, the cause of astrology's decline.

He must weep within his grave, but, for all his joy in his craft and conviction that astrology can be practised only with religious faith, Lilly's personal convictions, the idiosyncrasies of his persona, bred of time and place, betrayed him. Though his aim was quite the reverse, his writings and his influence played their own small part in building the gallows on which astrology was to be hanged.

PICTURES OF LILLY

We shall look here at William Lilly and his background in more breadth than our previous article, where we concentrated on his political attitudes. No student of horary can do better than to concentrate his studies on Lilly's masterly text-book, *Christian Astrology*, and as we have found our own studies greatly facilitated by some understanding of the man and his times, we believe attention directed to correcting some of the erroneous images thereof will not be wasted.

There are many pictures of Lilly in common circulation. The prevalent one, of course, is 'Lilly who?' The brief answer to this is that he is, by all evidence, England's greatest astrologer, whose spirit dwells happily among these very pages while his body lies beneath the choir-stalls of St. Mary's Church, Walton-on-Thames, where it turns violently in its grave every time someone mentions Chiron or links the eighth house with 'transformative experiences'. We shall return here shortly.

The favoured picture among those who have heard of him is of an austere figure who was leading astrologers through the wilderness when he went up a mountain and came back down with the Laws of Horary inscribed on two stone tablets – 'Thou shalt not not judge a chart with less than three degrees of a sign on the Ascendant,' and so forth. Which has inspired a large number of people who clearly have nothing better to do to spend a great deal of time and passion arguing about exactly what is written on these stone tablets. Woe betide anyone who dares to disagree with their conclusions!

A third picture, quite as unhelpful as either of these, is of a man just like us. We are lulled into this by the seductive democracy of the library: after the first couple of paragraphs, through which we are still aware that we are reading something old, the text becomes timeless, entering that bloodless limbo of the illustrious dead – or, more accurately, it becomes distorted into our Twentieth-Century mentality. We forget that Lilly lived and wrote blissfully free of such mental pollutants as Darwin and Jung, and in an age when thought was still regarded as something other than the rationalising of emotional responses. Just like us? No, he most definitely was not.

Then we have the academics' picture. They start with the assumption that astrology is obviously rubbish, so an astrologer must be a person of dubious morals for attempting to gull the public by practicing it. On this foundation, they construct an image of a Rasputin figure, knowing exactly what was going to happen years in advance (how he knew this if astrology is rubbish is a bit of a grey area), but in his machiavellian way releasing this information only in dribs and drabs as it happened to suit his financial or political interests. Wonderful indeed are the structures they create to close the circle of their arguments.

The Golden Age?

Lilly lived from 1602 to 1681, times of great turmoil in England. He had a reputation for accurate, predictive astrology which stretched across Europe. The words 'accurate' and 'predictive' are to be stressed, because his reputation was on the line all the time. Lilly wasn't in the business of telling people they were more sensitive than their partners realised, or had unfulfilled creative potentials. His astrology was hard, concrete and provable. So if he had this reputation, which he did, we must conclude that either our ancestors were too stupid to work out who had won a battle or whether someone was alive or dead – or that Lilly was rather good at what he did.

The years in which his practice flourished are commonly regarded as the golden age of English astrology, posing the riddle of how this Golden Age, when astrology prospered so highly, was also its death throes; for only fifty years later astrology was in much the same parlous condition that it is in today. But it seems probable that this idea of the mid-Seventeenth Century as a Golden Age is at least exaggerated, if not quite false. The difficulty is that this idea comes through the written word, and those who traffic in this coin – bless their dusty little hearts – have a touching belief that the people of real importance in the past are others of their own tribe, so if anything of real significance were going on, someone would have had the decency to write it down.

With no written record it is, of course, hard to establish what was happening: we can easily create idyllic fantasies about the past when there is little evidence either to work on or to contradict us. But the main reason that the mid-Seventeenth Century seems to be this astrological Golden Age is because so much written astrological evidence survives – and the reason for that has nothing to do with astrology. For a brief period, there was an almost total suspension of censorship, resulting in an avalanche of printed texts. Based on the amount of published material, the mid-Seventeenth Century seems to be the Golden Age of just about everything. Political historians are faced with their own avalanche of radical political and social writing. It is possible that this wealth of sophisticated radical argument appeared from nowhere, but this is unlikely: far more reasonable is the assumption that what was now being being printed was what had previously been spoken. So with astrology: it is possible that there was a sudden great flowering, which happened to coincide with the years without censorship, but it is unlikely.

We may consider the secondary sources. John Dryden, a contemporary of Lilly, was an enthusiastic astrologer. We should then expect to find astrological reference in his work, as astrology is part of his mental framework. Reference there is. But Shakespeare and, even further back, Chaucer, writing some 300 years before this supposed Golden Age, not only use astrological reference, showing that they themselves are familiar with the concepts, but they assume a sophisticated knowledge of astrology in their audience – as much, or arguably rather more than, Dryden and his contemporaries. This knowledge that their audience possessed must have come from somewhere: it wasn't gained from sun-sign columns in the daily papers. This is speculation and debatable – Chaucer's audience was an exclusive one – but it seems reasonable to suggest that the age of Lilly was, if anything, a silver age, the final flourish before decay. Quite possibly not even that.

The age.

Before looking specifically at the life of William Lilly, it is worth emphasising that he inhabited a world utterly different from our own. There are few places on earth today that are as different from our own experience as the world that he knew. Lilly did not unwind after a hard day at the horary mill by phoning for a pizza and watching TV. This is of course obvious – but for only as long as we deliberately keep it to the front of our mind: we tend naturally to slip into a vague background assumption that everything went on much as it did today. In some ways it did: many of the basic human concerns have not changed, so Lilly answered horaries on 'Does he love me?' and 'What can I do to earn a living?' There was a greater emphasis on some matters – 'Does she have any money and can I get hold of it?' – and less on others, as people didn't have to pay to be told 'No, he's not going to leave his wife;' but the physical, mental and spiritual world in which these questions were cast was not as ours.

To cite a couple of examples representative of this fact: most people would go to bed as soon as it got dark. We have a romantic picture of our ancestors spending their evenings sewing and singing psalms by candle-light; but candles were far too expensive for most to have in daily use. Even tapers were pricey, as well as being inefficient. One could go either to bed or to the inn, and navigating one's way home from there could be even more complicated than it can be today. Our second example concerns the image we have of people in the pillory or the stocks being pelted with rotten tomatoes by rosy-cheeked urchins. They weren't: they were pelted with the excrement that streets full of animals ensured was always in much more plentiful supply than tomatoes. Sources suggest that the favoured projectile – either for pelting people in the stocks or for throwing into the windows of rich people's carriages – was the dead cat, stock-piles of which seem to have been available on every street corner.

Lilly's practice would have been quite different from that of any western astrologer today, though having strong similarities with the norm in India. Entering his consulting room as he clocked on for work (no clocks in his house, and the town clock would not have told the minutes) he would have found a queue of people waiting for consultations. He would have set a chart for the day, which he would then have adjusted from time to time as necessary. Lilly was not overly troubled by niceties of precision. So long as his planetary positions were correct within a degree or so, he wasn't much fussed: an example to us all. On occasions where he felt an exact chart was needed, he would send it out to be cast by someone lacking his taurean unconcern for intricacies – a situation not dissimilar to the one we know, where the computer sets the chart and the astrologer judges it.

Dealing primarily with horaries, he would have spent around fifteen or twenty minutes with each client. This would include knocking the chart round to (approximately) the current time; listening to the situation and cajoling the client into phrasing the question in some tolerably coherent fashion; possibly telling the client the whereabouts of their warts and scars, as a convincer; and finally judging the chart. The reasons he could do this so quickly were partly the amount of practice he had, which was, by modern standards, enormous; but more because of perhaps the most significant practical difference between the astrologers of the past and those of today. Lilly and his peers were professionals – professional not merely in that they charged for their services, but in their whole approach to their practice. This is one of the razors which we must apply to our image of Lilly to cut it down to the truth. He treated astrology as a serious professional calling in a way that few if any of his descendants follow. His attitude to his clients reflected this approach to his craft: he turned them round quickly, providing them with concrete information: 'You want to know X? – OK, here's the answer. Thanks for your money. Goodbye. Next please.' One important consequence of this is that it becomes quite impossible to imagine Lilly looking at a chart and saying 'Oh dear! Only two degrees rising – I can't judge that. Put your money away.' This did not happen.

He charged a sliding scale of fees. It cost a great deal for a rich man to have a consultation with the famous Mr. Lilly; little or nothing for a poor one. To some sentimental minds this shows the man with the heart of gold. With such heavy emphasis on a Taurus second house, it may be more the practical realisation that there is no point in trying to charge the poor lots of money, because they haven't got it. The soda customer today may have been the soda customer tomorrow; but at least he was still a customer.

The life.

The problem with the autobiography, our main source for Lilly's life, is that it was written after the restoration of the monarchy, when Lilly's execution seemed more than likely. It is largely an exercise in proving that he had never done anything remotely reprehensible, which was a hard corner to fight. His lengthy explanation of exactly what he was or was not doing at the execution of King Charles is a remarkable attempt to drown guilt in a sea of fog. Interesting though it is, for a more accurate view we suggest a reading of Christopher Hill's *Milton and the English Revolution,* which despite mentioning Lilly only in passing, says more about him than any other book. While we cannot take one man's life as another's, fleshing out the bare bones of fact with the attitudes of Hill's *Milton* will give us a picture that is close enough to the truth.

Briefly, Lilly was born in Leicestershire in 1602. His parents falling on hard times, he walked to London for work: not quite the romantic Dick

Whittington picture of the youth with all his belongings in a spotted kerchief, wandering along the hedgerows; but trudging beside the cart that carried his belongings in a way cheaper, as fast and possibly less uncomfortable than actually riding upon it. He never worked as a scrivener, as he emphatically declares, but was a high-grade servant, one of his first duties being to perform a mastectomy on his master's wife. This operation was carried out in stages, but failed to arrest the cancer from which she died. His master remarried and then died himself; Lilly married the widow, thus achieving the financial security to spend his days playing bowls, attending sermons and pursuing his new-found interest in astrology.

In this brief picture of his early life, one significant fact has been omitted: Lilly's taking three years out on his way to London to gain a university degree. It has been omitted because it didn't happen – a point of the utmost importance in our understanding of his work. Most of what is written on Lilly is, by the nature of those who write things, written by people with an academic background. This is not necessarily helpful. The tendency to treat the astrological writings of Lilly as if they were a lost volume of T.S. Eliot and subject them to the kind of analysis deemed suitable for such causes nothing but confusion. Lilly was not an academic, and should not be treated as if he were.

Any reader who has learned a craft will be familiar with a common situation in training. The craftsman is working busily. The apprentice asks, 'What are you going to do about that there?' to which the craftsman replies, 'Pass me the whatsisname; I'll give it a bit of how's your father and Bob's your uncle.' Unlikely as it may seem to the uninitiated, 'whatsisname', 'how's your father', and 'Bob's your uncle' are precise technical terms – which can mean absolutely anything, depending on the context. But in the situation their meaning is quite clear to both master and apprentice. The academic who constructs an analysis of the master's use of the term 'whatsisname', comparing its meanings in different situations, is going to tie himself and his readers into all sorts of unproductive knots. Yet this is exactly what we see in so much modern writing on traditional astrology.

Astrology is a craft. That is, it is a hands-on working in the real world; and because it deals with the real world, it doesn't correspond with the tidy, abstract rules of grammar by which the academic attempts to render reality explicable. Lilly was a craftsman, and he wrote as such. We would save ourselves a deal of ink and a deal of bad feeling if we remembered this.

Lilly began seeing clients in about 1635. In 1642, Civil War broke out. The hurly-burly was not done until the mid-1660s, when the monarchy had been restored and a satisfactory degree of vengeance taken. Lilly's life must be seen against this backdrop; without some understanding of what was going on, many of his actions make little sense.

Hugh Trevor-Roper claimed that there were no problems at the start of the war that could not have been settled by a group of men sitting around a table. If this group would not have had to include King Charles, this might have been true – if the problems could have been handled one at a time; but they came in their battalions. There were massive strains at all levels of society as the economy was realigned to the prototype of what we have today, and a powerful ground-swell of revulsion at the debauchery and decadence of the court. Although the moral tone had improved somewhat with the accession of King Charles, the damage had already been done. Most significant of all was the intertwining of religion and politics in a way that is incomprehensible in the West today – far beyond anything we might see in, for example, Northern Ireland. Political and religious radicalism went hand in hand; more precisely, political radicalism was conceived in religious terms and religious radicalism had what were regarded as inevitable political consequences. Lilly was deeply and passionately involved in this. He was one of Parliament's leading propagandists. On the upper decks, Milton fought the intellectual war with the leading thinkers of Europe, justifying the cause; down below, Lilly fought the popular war by demonstrating that these changes were divinely ordained and inevitable, as shown in the stars.

Our view of the conflict is heavily coloured by the TV dramas we watched over Sunday tea. The Cavaliers had long hair and fancy clothes, while the grumpy Roundheads closed down the theatres and were indisputably the bad guys: a picture more romantic than accurate. The Cavaliers may have worn pretty clothes (as did most of their opponents), and to deny them the better cause is to deny the truths at the heart of our astrology; but the prevailing interest of too many among them was the lining of their own pockets. The Roundhead leaders were rather better behaved, but equally self-serving. A high proportion of the Parliamentarian rank and file, however, was fighting for a political and religious ideal. Lilly was as idealistically committed to this cause as any.

Our picture of this Puritan cause is again a distorted one. Our image of the Puritan has much more to do with late Victorian non-conformity than anything recognizable from the Seventeenth Century. The common idea is of someone like Hudson the Butler in *Upstairs Downstairs.* If we remember that one of the beliefs shared by several of the Puritan sects was the urgent necessity of free love, we begin to see the flaws in this picture: Hudson the Butler wasn't big on free love. As for the closing down of the theatres, this was not a moralistic wet blanket, but an act of political pragmatism: as the theatres were centres of dissent, to have allowed them to stay open would have been suicidal. As is clearly shown in his writing, Lilly, a committed Puritan, was no more the sour-faced spoilsport than were either Cromwell or Milton.

Exactly where Lilly stood on the broad platform of beliefs that was Puritanism is unclear. On a scale of radicalism from one to ten, he probably clocked in at around the 6 or 7 mark: he had no time for the ranters, extremists whose favoured recreation was tearing off their clothes and grinding their teeth in the windows of rich people's carriages, but did believe firmly in the approaching millennium – the coming of Christ's kingdom on Earth, for which end earthly kings must first be overthrown. Those who ordered the execution of the King did so from the conviction that by so doing they were furthering the cause of the saints; if we compare the enthusiasm with which Lilly cheers this cause with the coldness of the autobiography in which he claims to have done no such thing, the truth of his feelings is evident. There is a cosy belief among the moderns that Lilly hated the monarchy but thought the King a decent enough chap. This is untrue. To accept it means foisting completely anachronistic ideas onto Lilly. He would have regarded this sentimental humanism as a betrayal of faith: he was not a Twentieth-Century man.

Lilly, in his own way, was a soldier for this cause, as involved as any in the line of battle. His almanacs were the best-sellers of the day, a favourable prediction from Lilly being said to have the value of a battalion of soldiers to the Parliamentary armies. Lilly was not impartial. Far from it: he was deeply engaged with that strange stuff that proceeds outside astrologers' windows – the mysterious business called Real Life. His astrological writings were not composed for the inhabitants of any ivory tower. They were his utmost contribution to ushering in the rule of the saints and the second coming of Jesus Christ on Earth. Astrology without engagement is a waste of time: it becomes nothing but a glorified crossword puzzle. Lilly never aspired to the dubious ideals of academic abstraction of which he is accused of falling short; his great achievement lies in managing to be engaged without the compromise of all astrological credibility that we find in his royalist rival, Gadbury.

A mere ten years after Charles I was executed, the brave new world was falling apart. The Revolution had been killed off quickly: immediately victory was theirs, the Parliamentary leaders explained to the ardent rank and file that they hadn't really meant the idealistic rhetoric, shot the cadres and packed the rest off to fight the Irish. By 1660, the least bad option seemed – even to many of those who had helped pull down the monarchy – to be the Restoration. This was devastating. The dream had failed. What was worse, it hadn't failed through being beaten in the field: it had fallen apart through its own internal faults, through the inadequacies of those who had carried it. So much had been promised: the unprecedented event of a king tried by his subjects and then condemned; the execution, so staggering that 'women miscarried, men fell into melancholy, some with consternations expired'; and then – nothing.

Here we see the cause of Lilly's gradual withdrawal from predictive astrology. We must remember that he had seen these as the Last Days – that is, as an ordered part of history. Far more than the crushing sense of failure felt by anyone who has fought for an ideal and lost was the failure of (his understanding of) divine order, which is of course, the basis of astrology.

Medicine had always been an enthusiasm; now he concentrated ever more upon it, in 1670 finally obtaining – after much political wrangling – a licence allowing him to practice officially. This was not an uncommon path for failed revolutionaries either then or since: the medical profession has always provided more than its fair share at the barricades,[1] and Lilly was just one of many idealists who turned to the immediate practical help they could offer through physical healing. One of his most spectacular astrological successes was yet to come, but his prediction of the Great Fire of London in 1666 had been made long before in his publications of 1648 and 1651. Although still working with astrology, a combination of caution in a hostile world and the wisdom

[1] A contributory reason for the success of the New Model Army was that it had many more doctors and surgeons than the Royalist armies. This not only helped in physical terms, but had immense benefits in morale, as the troops felt they were being well looked after.

that is found in that withered field where the farmer ploughs for bread in vain had changed his focus from the public to the personal. By this time he had buried his second wife and married someone he actually liked – Ruth, with whom he had a long and happy marriage. His time was spent quietly, treating the ills of the populace of Hersham – often for free, which was itself considered a revolutionary act – until he died in 1681.

That his nativity was published by his enemy, Gadbury, has raised doubts about its veracity: in an astrologically literate age, a common means of attack was, rather than vilifying one's foes, to adopt the subtle method of publishing a plausible but unfortunate birth-chart, relying on one's readers to draw their own conclusions. The chart is its own advocate, and the internal evidence is such that it must be accepted. More even than that painting of his notable contemporary, it reveals the man 'warts and all'; for all his failings, however, we lack his peer, and in the workshop a kettle is forever boiling on the hob and a cherry-cake kept freshly sliced in case he should drop by.

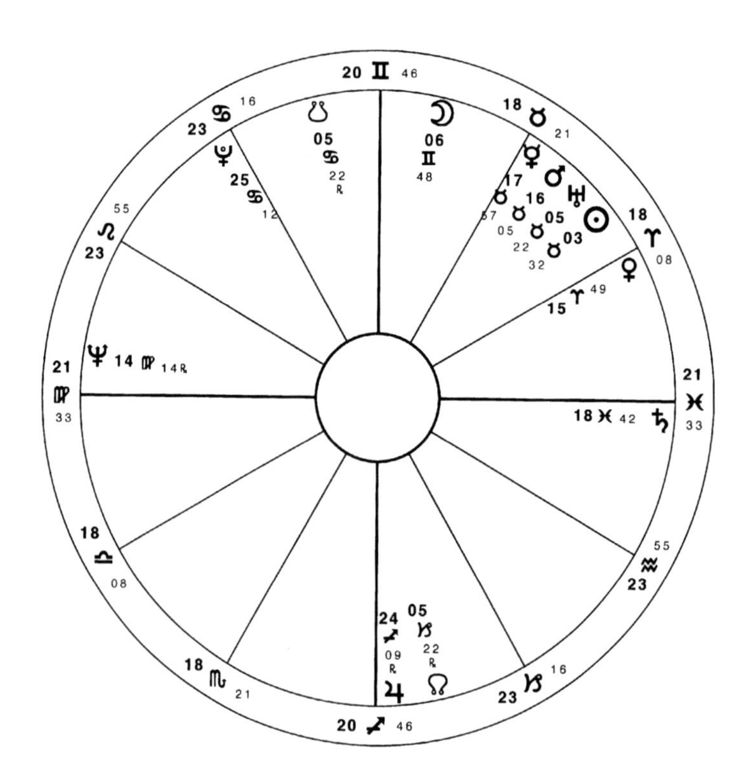

Chart 8. William Lilly, nativity. May 11 1602 NS. 2.04 am LMT. Diseworth.

NEPTUNIA REPLIES...

– a word from our sensitive seer

Dear Neptunia,
I am so desperate. Whenever I do horaries, I end up with only a few scribbled notes on a scrap of paper. But my boyfriend has started using the special *Lillywhizz* computer program for horaries. This presents him with seventeen bound volumes of planetary information, neatly tabulated and indexed, for each chart he casts. Now he uses it all the time. When I see the delivery lorries roll up outside his door to drop off all the information this program produces, and think of my poor scrap of paper, I feel so inadequate. What can I do?
Yours in desperation, Tracey.

Dear Tracey, Only the other week, my flame-red Lamborghini was running low on petrol, so I decided to change it. As I wandered round the show-rooms in search of a suitable replacement, I was assailed from all sides with assurances that this model's computerised nose-scratcher and that model's passenger-side swimming-pool are just what I need to carry me from A to B with a reasonable certainty of arrival. But, Tracey, they are not. That which I need is simply an engine, four wheels and a modestly attractive body capable of holding me in and the weather out. Indeed, the integral multiplex cinema, shopping-mall and Michelin-starred restaurant (fitted as standard on model G4i, optional on others) is more likely to hinder my desire to arrive quickly and in one piece by distracting my attention from the business of driving. It is only thus with your boyfriend's *Lillywhizz.*

Do you remember your school days, and the horror of exams? Between you and me, Tracey, I was able to pass French only by having one of the grooms sat at the desk in front of me, with the full conjugations of several irregular verbs tattooed on the back of his neck. No doubt your class-mates, and perhaps even you yourself, were like mine, seduced into buying piles of glossy Pass-notes and Exam-aids by the wishful illusion that just owning them would somehow insert the knowledge into their brains. But it did not, and those who passed were invariably the ones who had taken the trouble to learn those damn verbs by heart.

It is so tempting to think there is an easy way to knowledge; that it can be bought in a book or a computer program. It can't. It can be bought only with the coin of effort and experience. It is so easy to go to the shop, rather than spend another hour poring over Lilly; and your boyfriend's program, I am afraid, is a way to avoid making the effort that might turn him into an astrologer. Moving hand to pocket can never substitute for moving brain to chart.

Lining one's shelves with books does not make one a better astrologer; lining one's head with experience does. I have told you before, Tracey – horary is like surgery: get in, get out and get the body sewn up as quickly as possible. By the time your boyfriend has picked his way through the avalanche of irrelevant information given by *Lillywhizz,* his patient will have bled to death. When you are performing a tonsillectomy, it is not necessary to spend time examining the state of the patient's appendix.

You stick to your scrap of paper, Tracey, and be proud of it. To paraphrase that nice Mr. Culpeper, we would all be better off for keeping our wits in our heads, for that's the place ordained for them, and not in our computers.

Your caring, Neptunia

4

The Background

THE MOST BEAUTIFUL MUSIC

Finn McCool and his companions were out riding one day, hunting the wild boar through the wooded hills of Ulster. While they rested at midday, lying eating in the sunlight of a forest glade, McCool posed the question, 'What is the most beautiful music of them all?'

The fearsome, one-eyed warrior, Golla MacMorna, spoke first. 'It is the sound of battle,' he said. 'The sound of sword on sword, of the spear in flight; the sound of fear and of victory.'

Then spoke Diarmid, so beautiful that no woman could look on him and not lose her heart. 'It is the sound of a soft voice calling from her chamber in the night; the sound of sweet words whispered in the dark; the faint trembling of lips as they hover for that first long waited kiss.'

Then Fergus spoke, who told of the singing of the wind through the cornfields near his home; Connor, of the tympani of waves crashing on the shore; Conan, of the murmur of his child in sleep; and Oisin, Finn's own son, of the warmth and wisdom in a father's voice.

Each one answered, each with his differing view. Then, when all were quiet, Oisin asked, 'And my father, Finn McCool: what say you is the most beautiful music of them all?'

'The music of what happens,' said Finn McCool; 'That is the most beautiful music of them all.'

And that is what, as astrologers, we are privileged to study: the music of what happens, indeed the most beautiful of them all.

There are many ways in which man has attempted to make this music intelligible – to read the score, as it were. Some of these are inevitably more successful than others. The experimental methods of what is now, for some reason not immediately obvious, called 'science' seek not to read the score or hear the music, but to understand it by examining its effects on its listeners, the existent animate and inanimate entities of the world, so putting many levels of opaque reality between themselves and the composer. At the other extreme, the mystic attempts to comprehend by realising his oneness with the mind that is creating this music.

Of what might loosely be called the divinatory arts, though limiting astrology to this does her a great disservice, some attempt to predict by humming along with the tune until the operator, if skilled enough, can catch sufficient of its form to gauge where it is going next, while some, of which astrology is the epitome, use the vestiges of true scientific method to objectively – or dis-involvedly – understand the nature of the forms from which the music is built: its notes and tempi, for example. From an understanding of these forms – the building blocks of the music of what happens – the astrologer can then proceed in two directions: to understand the music that is made from these blocks and thus, if he wishes, predict its flow, and to understand the mind that created the blocks. The astrology that we have is, in this sense, a fragment of a structured, disciplined mystical science.

Plotinus says that if we establish the comprehensive principle of co-ordination behind all manifested phenomena 'we have a reasonable basis for the divination, not only by the stars, but also by birds and other animals, from which we derive guidance in our varied concerns.' That is, if we imagine all manifested phenomena as two dots on the surface of a balloon, these dots will move as the balloon is blown up. It is not until we realise that the balloon is being blown up and that this has an effect on the dots that their movement becomes comprehensible to us. Once we have grasped the basic coordinating principle of the balloon's expansion, a knowledge of the movement of one dot will enable us to determine the movement of the other. If one of the dots is me, it is of no matter whether the other dot is the planet Venus or my cat: the understanding of the basic coordinating principle will still enable me to deduce things about my own position from observing it. Over the centuries, the position of the planet Venus has proved easier to tabulate.

In practice, of course, the position is rather more complex than the metaphor suggests, in that we have the familiar Aristotelian principle of balloons within balloons; but the idea remains the same.

It is the size and apparent regularity of orbit of the planets that has made them of so much more practical use than the movements of birds or animals, especially for a sedentary race increasingly removed from contact with the natural word against which the movements of animals must be seen if they are to become comprehensible. In India, the classical model of the astrologer at work has him seated in a clearing, making judgement from the surrounding world as well as from the chart itself:

the weather, the direction from which the client comes, his clothing, movements of animals, the chart – all are used as one.

That we are a sedentary and, increasingly, an urban race has a profound effect on our choice of technique for grasping the coordinating principle. We judge from pieces of paper rather than the livers of newly-slaughtered sheep; but the form, too, of our astrology has been shaped by our culture.

Although this culture has become so universally embracing that we almost forget the existence of an alternative, one of the most fundamental divisions of humanity is into nomadic and sedentary peoples. In *Genesis*, only the divisions into man and woman and parent and child come before the division into Cains and Abels. Just as man and woman or parent and child, nomad and settler have profoundly different views of the world: a nomad is not just a settler who moves about from time to time. The nomad would find the concern of Scarlett O'Hara for Tara quite incomprehensible: indeed, in the friction between Rhett Butler and Scarlett, we can see something of the difference between nomadic and sedentary values, and the power of that perennial romantic attraction which the nomad holds for the settled soul.

For the settler, place remains fixed while time is ever moving. The arts of the sedentary cultures fix time in place, painting or sculpture being examples; cinema or TV being the most typical today. The settler's dream is to found a 'house', in literal or metaphoric sense, that will be passed on through generations. For the nomad, time is his static medium, one year being like the next with none of the gradual sense of growth with which the farmer is familiar, while place moves, the horizon moving always before him. The nomad's arts move through time: poetry, which has no place, but which starts at one time and finishes a certain time later, and song are his chosen media. The nomad's dream is to have a name that will be passed on through generations, to achieve a famous deed that will be recited by the bards. Note the opposition when we phrase this in astrological terms: the settler's dream is the fourth house; the nomad's the tenth.

Note also the significance of which of these is above the horizon: the nomad's deed must be honourable, done in broad daylight, for all to see, for nothing is worse than to remembered for the wrong reasons; but for the settler, whether his house is achieved in honourable or dishonourable ways is largely immaterial: in the darkness of the IC, the foundations of his house cannot be seen. Whatever dark deeds may have built his house, they will be washed away with time. This difference in value is the basis for much of our literature, cinema and, indeed, political thought.

Nomad and settler have their different astrologies. The chart is a mixture of time and place: a certain time, at a certain place. The nomad uses whole-sign houses, which throw the emphasis of the chart onto the time. The sedentary cultures have developed the myriad house-systems (the choice of the word 'house' for a section of the chart is not coincidental), stressing the place. The term 'whole-sign houses' is in fact a settler's mistranslation of what the nomadic astrologer is using, for he has, being a nomad, no houses: he does not think 'If the Ascendant is in Gemini, the second house is Cancer,' but simply 'If the Ascendant is in Gemini, the second sign is Cancer – and the second sign is concerned with possessions'.

Apart from anything else, whole-sign houses are simple to calculate while whipping one's sturdy little pony across the endless steppe. I have yet to come across anyone who can work out Regiomontanus cusps from scratch (no books of tables if you're a nomad!) in their head. The whole paraphernalia of our astrology is specific to our culture.

The colouring of astrology by culture did not stop with the split between nomadic and sedentary peoples. Far from existing in a world of intellectual purity, our astrology has been moulded by our changing social world (no less than the modern pseudo-sciences of physics or biology). We could doubtless find a cultural concurrence to many of the astrological divisions: those who reverse Fortuna by night and those who don't, for example. The most obvious division is that which occurred with the Renaissance.

The Renaissance, a gradual process extending over several hundred years rather than the sharp watershed that its name implies, saw a reversal of man's priorities as total as the reversal in the model of the solar system which is regarded as one of the great achievements of that age. Our position, dwelling in the foothills of that mighty mountain, makes it impossible for us to judge clearly its effects: its shadow falls across all our thought, however hard we may strive to escape it. But for all its pervasive influence, it is possible to see that it was not necessarily the unreservedly Good Thing that Whiggish attitudes to history assure us it was. Just because it brought us to here does not necessarily mean that here is the best place to be.

The discoveries of the Renaissance are manifold, from double-entry book-keeping to America, and, like all discoveries, they each reflect upon the age that made them: without the mental or spiritual capacity to receive that discovery, the discovery will not be made – we might evince the Vikings' stumbling across America as a discovery made before the capacity to receive it was formed. Perhaps the one that most clearly symbolises the spiritual transformation of that age was one of its favourites: that of linear perspective.

To us it is self-evident that a drawing made according to the rules of linear perspective looks 'real', while the typical product of a medieval artist does not. Yet it is only because we have lived through the Renaissance that we experience this: only because we have accepted the Renaissance reversal of values, selling our birthright for a mess of sweet-smelling but nutritionally empty pottage. That it bears a resemblance to the superficial form of an object does not make something look real; this is the error of our contemporary scientists, with their conviction that if you can measure something you understand it. As astrology teaches us, the reality of an object is in its essence, not in its form; throughout history, it is only cultures in their decadence that have produced superficially naturalistic art: cultures still in touch with their heart produce art that concerns itself with essences, with the spiritual core. Naturalistic art may look like the surface of a form; this is not the same as looking real.

The significant thing about linear perspective is that it is made to be viewed from one particular spot, unlike a medieval work depicting essential truth which looks 'real' from whatever viewpoint it is seen. That is, linear perspective prizes the individual viewer in its abandonment of spiritual for mundane value. As astrology is always coloured by the society in which it lives, this is exactly what happened to astrology at the same time.

The correlating change in astrology during the Renaissance was the switch in emphasis from horary and mundane astrology to natal (though natal existed before and horary since). The point of horary is that the artist is divining the Will of God on the particular issue under discussion. The form of the question is always, although rarely stated as such, 'Is it the Will of God that X or Y...?' It is possible, too, to treat a natal judgement as a divination of the Will of God for that life; but the way of the world being such as it is, natal astrology inevitably descended into what it is today: not a divination of the spiritual role of the life, but a revelling in the idiosyncrasies of personality, the very things that draw us away from the spiritual.

Apart from the contemporary role of astrological consultation as titillation, providing the otherwise rare opportunity to talk about nothing but ME for an hour or so, this difference in emphasis is perhaps clearest in the the dramatic reversal of attitude towards death in the horoscope. For the ancients, when judging a nativity the first thing to do was to work out when the native would die: without this, any further judgment was clearly meaningless. But this was in an age where the spiritual verities were accorded more importance than today: in spiritual terms it is the moment of death that gives the life its meaning. In secular terms, the moment of death is the extinction of the ego, and so, as modern astrology exists solely for the amusement of that ego, the textbooks now agree that one must never suggest to one's clients that they might possibly be mortal.

It is this reversal of value between the spiritual and the profane, as total as the reversal in an electric current, if happening over a rather longer period of time, that explains the Galileo controversy. Contrary to what we are taught, this was not a matter of a true view superseding a false one: the victors write the histories, and this picture is painted by the world which the Galilean view has created, a world whose unacknowledged legislators are the scientists whose conviction it is that man does live by bread alone.

The Galilean controversy was not a question of true versus false, but of which level of truth was to be regarded as of prime significance, the spiritual or the profane. It is this that caused the opposition of the Church to Galileo's teaching. Throughout the Middle Ages, the Church displayed a notable openness to new ideas: this catholic approach to knowledge was not suddenly abandoned without reason. The Church's opposition came from an astute awareness of the spiritual and moral dereliction that the Galilean model would inevitably draw in its train, being both product and cause of a turning from true to illusory value. It is its importance as a spiritual rather than purely intellectual phenomenon that caused the two-hundred year delay in this model being widely accepted: culture moves at a slow pace. (Consider that it will doubtless be another couple of hundred years before relativity becomes widely accepted as anything other than an intellectual abstraction. As Gramsci pointed out, 'One can think what would happen if in primary and secondary schools sciences were taught on the basis of Einsteinian relativity...the children would not understand anything at all and the clash between school teaching and family and popular life would be such that the school would become an object of ridicule and caricature.' Such was the shift in commonplace value demanded by the Galilean model.)

So our study of astrology has two possible levels, two perceptions of truth between which we must choose, which perceptions might almost be described as the pre- and post-Galilean. We may listen to the music of what happens, or we may listen to the raucous dance of the individual ego. Seen in this light, the refusal of the traditional schools to admit the outer planets to judgment is in perfect decorum, because in the cosmological model that is relevant for the perception of truth in the spiritual sense, they do not exist.

That they have now been discovered does not make them any more relevant to our model than are Nintendos or barbecue-flavour crisps. They may be there, and they may be useful if our aim is the titillation of the ego, but they are not part of the model of an astrology that is a true science: a path to knowledge of the Composer of the most beautiful music of them all.

TIME IS MONEY

Fundamental to astrology is the idea that things happen at specific times. Not random times, at which these events chance to have fallen; but definite moments for which alone they are fit. Modern science has a similar idea; although, in common with all the ideas which it has taken from the Tradition, it has it only in debased condition (let us dismiss once and for all the strange notion that the scientists are somehow rediscovering the truths of which the great faiths have spoken: the man on the down escalator does not reach the top, no matter how far he travels).

The most refined manifestation of this idea in modern science is as part of the Whiggish concept of progress, by which all Good Things in the past have been stepping stones to the peak of excellence at which we have now arrived. In more basic form, we have the common-sense statement that one thing must happen before another which depends upon it becomes possible: you don't invent the remote-control until you have invented the TV. So it is with Mr. Darwin, whose ideas are such a milestone on the road to the Dark Age in which we dwell. Both the traditional and the scientific point of view agree that his ideas could have been distilled only at a certain time. He of traditional bent will hold that only when the prevailing state of reason reached sufficient stage of corruption could ideas like his be taken seriously. His scientific antagonist will hold that certain base-camps had to be established before even Darwin's towering intellect could scale the Everest of evolution. This is something of a circular argument, but we shall let that pass.

One of these vital base-camps concerned money. A certain stage in the progress/corruption of monetary thought had to be attained in order that Darwin might stand upon it to take his great leap forward. This stage was the introduction of paper credit.[1] This happened during the early part of the nineteenth century. Simply, this enabled economists to produce as much money as they wanted just by writing a few noughts on

[1] See Martin J.S. Rudwick, *Poulet Scrope on the Volcanoes of the Auvergne,* Brit. Journal for the History of Science (1974) Vol VII, for more detail than space here permits. (As traditionalists, of course, we are not able to increase the size of the magazine by adding a few more noughts).

a piece of paper. Darwin's particular problem was time. For his theory to have even that degree of plausibility that it does, he had to vastly extend the amount of time Nature had available for the transformation of microbes into men. Instead of the mere few thousand years that was the prevailing estimate of the age of the Earth, Darwin needed millions. The advent of paper credit showed him how this could be done: simply write a few extra noughts and, hey presto, you have it. This points a connection between time and money which is not without importance for astrologers.

The connection between the two things is easily seen: we spend, waste or save them both. But let us go back to the beginning. As our immediate measures of time are given by the Sun and the Moon, so were these once the mainstay of the monetary system. There was the odd jewel in circulation, which we might equate with a planet, but the pound in your pocket was of either silver or gold.

There was a qualitative difference between the two. If you wished to buy my goat we could agree a sum in either gold or silver; but these piles of metal which could purchase the same things were most surely not the same. This becomes clearer if we draw in the third strand of our argument: intellectual history. The Sun and the Moon are intellectual and rational knowledge respectively. Just as the Moon will help you keep to your path when the Sun has set, rational knowledge can keep you focussed when your powers of intellection are on the wane. But no amount of rational thought is equal to one moment of intellection. Think as much as you like, it remains thought and will never become knowledge. We are reminded of Thomas Aquinas, realising that the matchless achievements of his reason were worth only so much straw. It is indeed symbolic of the dereliction of modern thought that an intellectual is now taken to mean not someone who practises intellection, but someone who thinks a lot.

The Sun is divine and the Moon mortal; it is the immediate symbol of creation and, with its blemished face, the immediate reminder of our frailty – and of the frailty of our reason. It is not insignificant that the fee of traitors is traditionally paid in silver – 'For a handful of silver he left us,' – the appropriate wage for their human fallibility. As silver is the Moon and gold the Sun, we may pursue this logic: if twelve disciples equal one year, at a piece per day, one disciple must cost thirty pieces of silver.

At first, the objects of exchange had innate value of their own. You would present me with a finely wrought golden shield; I would reciprocate with a tripod for sacrifice. Here it is obvious that there is qualitative difference: no number of tripods equals one shield, and *vice versa.* The first stage of corruption was the introduction of coinage. In some way or another, ten silver coins is held to have the same value as one of gold. Coinage has turned the Sun and Moon from things of value into money, which is a system of measurement. Everything, we are told, fits somewhere onto this scale of measurement by money: that Hollywood may arouse our sentiments by claiming otherwise shows only how ingrained this idea has become. This may be all very well in quantitative terms – if my pig is worth two pieces of silver, both my pigs are worth four – but has been a recurrent source of contention when applied to the qualitative: we soon come against the old anarchist conundrum of the number of bricks a bricklayer must lay to deserve the surgeon's wage.

In the early days of coinage, when the material from which the coin was made was still of more import than the number stamped upon its face, debasement of the coinage was a sin equatable with heresy, and punished with similar severity. As the introduction of base thought into the creed threatened the Church, so the introduction of base metal into the coin struck at the heart of the state. The effect was like an eclipse: the Sun obscured by silver, or silver obscured by the shadow of the Earth.

The significance of the introduction of coinage is clearer if compared with the equivalent change in our perception of time. Coinage equates with ratiocincration (in period as well as meaning). Time, too, like our shield or tripod had once a value in itself. This is the perception of time on which astrology is based: the knowledge that each moment has its individual nature, that to every thing there is a season. Hand in hand with the introduction of coinage comes the spreading and now ubiquitous belief that moments have no qualitative difference, that time is a constant. This belief is very handy if time becomes something to be bought and sold – one's eight hours in the factory or the field.

But we soon discovered a practical problem with gold and silver: there isn't much of them. When people want them, it becomes pressing to find more. This is a quite different situation from that with which too many of our readers will be familiar: you may be short of cash, but there is plenty of it about – do some work, rob a bank, marry wisely and some of it will be yours. One of the greatest formative influences on the history of pre-modern Europe was the sheer shortage of specie.

That historian of civilisations, Braudel, claims that 'If medieval Islam towered above the Old Continent, from the Atlantic to the Pacific for centuries on end, it was because no state (Byzantium apart) could compete with its gold and silver money, dinars and dirhems. They were the instruments of its power.' Modern historians, viewing that age through glasses coloured by the fumes of this, see medieval Europe as a pond in drought, filled with fish struggling for the few remaining drops of water. Then it rained. Seen from this modern viewpoint, a viewpoint which would be quite incomprehensible to a native of that time, it was an age of restriction: limited specie; limited knowledge, as exemplified by books in small number copied laboriously by hand, while monks debated angels and pins; limited time, with the span of life so short.

Water precipitates in many forms: as snow, dew, rain or hail. So this rainstorm fell in its different forms. There was a shower of gold as the treasure-ships began to arrive from the New World. Within a very few years, Europe was transformed. There was a flood of words as the printing-presses sprang to life, diluting the intellectual in the rational – reflecting the effect of the mass of new coinage. Echoing from level to level of meaning, this new wealth financed the swing from spiritual to secular in the political sphere, a swing legitimised by the works rolling off the presses. With this flood of gold so strong, it is hardly surprising that the scientists decided the world must revolve around it. We have already considered what was happening to time at this time, as mirrored in the nice trick of linear perspective.[2]

Thence the invention of paper credit: the true product of the Age of Reason, as truth is pushed aside and replaced by fancy. The obvious contemporary comparison in the intellectual world is with the death of the solar art of poetry – the art of telling the truth – and its replacement by the lunar art of the novel – the art of telling tales. The main impetus behind the creation of the novel being to fill time, of which, with these brave new developments, there was now suddenly too much. As time was now a kind of coinage, it could be accumulated until people had far more of it than they knew how to spend.

The idea of monetary gold and silver has itself long been abandoned: no more Gold Standard; no more of the promise 'to pay the bearer on demand'; not much now of even paper money. Intellection is no longer a

[2] [See *The Most Beautiful Music*, in this volume.]

respected guest in the world of thought, and thought itself has been replaced by artificial currencies of no inherent value. As we might expect, the same has happened in astrology: the Sun and Moon are no longer the vivifying force in the chart. While they were once the planets' parents they have now been reduced to the status of slightly older brother and sister. And we too have our paper credit: when the astrologer finds himself unduly cramped by working with just seven planets, he can now have as many as he wants. 'New planet, sir? Certainly. Any particular colour in mind?' He no longer needs even to have his planets of base metal: they need not have any real existence at all. But with no gold and silver, no Sun and Moon, we have no truth.

The Sun's not yellow, it's a chicken

When currency has been debased it becomes harder to distinguish the real thing when we see it, because no one has seen it before; no one knows how it should behave or how to test it. You might try presenting a gold coin at Woolworth's, but you will need a pocket full of cupro-nickel if you wish to be served. The fairy-tales are familiar: the pretend princes all behave as we think princes should; the real prince is rejected for behaving as princes do.

This is true in the big world of faith; it is true too in our little world of astrological study. We don't know how to handle the Tradition when we find it. So unaccustomed are we to truth, we assume it is just a collection of opinions like everything else that we hear. So much that claims to be drawn from the Tradition reminds us of Gurdjieff's recipe for commercial chicken soup: put some parsley in a vat of water, then chase a chicken through the kitchen. Writing a book on traditional astrology? Switch on your word-processor and use a volume of Lilly to prop up your desk.

Rather than speaking from within the Tradition, too many of the modern writers on traditional astrology write as Twentieth-Century man, the monarch of all the intellectual realms that he surveys, picking a piece here and a piece there as if he had before him so many boxes of chocolates: a coffee cream from Lilly, a caramel crunch from Barbara Watters or even Alan Leo. These authors show a remarkable and lamentable readiness to dismiss ideas that they do not understand. We cannot come to the Tradition as conquerors, selecting which pieces may live and which shall die; we can come only as supplicants. The only correct attitude to the Tradition is that if we do not agree with it, we are probably wrong; if we do not understand, we should keep quiet until we do. Traditional knowledge is an absolute monarchy: it is not an elected government. In this monarchy the knowledge of Bonatus is not comparable with the thoughts of Ms. Watters: they are two different currencies, between which there is no rate of exchange.

These are ideas that fall strangely on contemporary ears, betraying the nostalgic longing of their writer for his happy childhood at the court of Genghis Khan. But there are no two ways about it: if there is validity in the Tradition, our modern process of thought is wrong; our currency is invalid. We were discussing some modern astrological theories with a respected astrological writer recently, when we had cause to mention one of the fundamental philosophical ideas behind the Tradition; an idea on which every one of the great faiths agrees. His casual response of 'I don't accept that,' would be no doubt unremarkable in most circles; but in the workshop it caused sudden consternation, and the stable-lads and even a couple of the more impressionable journeymen were quickly put about their duties, safely out of earshot. From the point of view of the Tradition, what a very odd thing to say. We do occasionally find the newer apprentices coming out with similar views; they are sent off on their own for a period of reflection until they realise exactly where their thinking is wrong. We do, of course, acknowledge the possibility that they may have climbed an intellectual peak that such as Aquinas, Ghazali, Ibn 'Arabi not only did not climb but quite failed to notice; but we know that the possibility of this is a small one. In our astrology, we may be dealing with craft knowledge rather than intellectual perception, but the principle is the same.

Implicit in the idea of Tradition is that the tradition can be meaningfully criticised only from within; if we wish to see whether the fan-vaulting is up to the mark, we must first enter the cathedral. It matters not what our personal opinion of the cathedral might be: we still cannot see the vaulting without going inside. To believe otherwise is to fall for what John Morrill, a leading historian of the Seventeenth Century, whose work is to be recommended to anyone wishing to understand the world in which Lilly lived, has called 'the greatest lie of all in the sciences and metaphysics: that we are the product of a process of maturation, in which all our knowledge is superior to that of all other cultures; and that we have refined ourselves out of and beyond most of the nonsense that held back previous cultures.' If we are going to study (and, even more, write about) the Tradition, we must get off our immobile Twentieth-Century behinds and step inside it first.

NEPTUNIA REPLIES

– a word from our sensitive seer

Dear Neptunia, I am so desperate – no one can help me but you! I had this lovely stone circle in my back garden, full of nice things like cushions and scented candles, where I would worship the Moon goddess. Then just last week my boyfriend came back from the pub, moved some of the stones around, stuck his smelly trainers right in the middle of it and said he had re-dedicated it to the Sun god. Then he sent me indoors to do the washing-up. I am so unhappy – what can I do?

Yours in desperation, Tracey

Dear Tracey, Yours is a common problem – a problem that goes back thousands of years. People build themselves lovely stone circles dedicated to a female deity. They enjoy an idyllic life-style, passing their days in nattering and the exchange of knitting patterns, and all is beautiful. Then one day a bunch of men arrives and tells them that everything has to change. The temple is altered and made over to a masculine, solar god, chocolate suddenly becomes fattening and there is far too much sport on TV. Believe me, if I have seen it once, I have seen it a thousand times.

But I ask you Tracey – have you really thought about what is going on here? Are you quite sure you're not forcing what you have seen to fit the peculiar conception of 'thought' that is current in this gender-obsessed society? It is common to look back over the millennia and find a wonderful golden (or should I rather say silver?) age when all was peace and plenty, which bounty was bred by humanity's predilection for female, lunar goddesses. This then came to an abrupt end as masculine, solar gods were installed. Adopting the most simplistic of attitudes, this is seen as a Bad Thing. This enables us to indulge in a comforting nostalgia for times past – if *only* we lived in a society where we were understood.

It is, you must, I think, admit, convenient that all these changes happened in prehistoric times, so we can twist what little evidence remains to fit whatever theory we are trying to prove. But rather than probe these suspect pictures of human society, I shall ask you but one question, Tracey. Has your boyfriend started staying at home more often in the evenings?

If you have noticed more of his presence around the house, even when he isn't there himself, you may well have unwittingly seen the cause of this change. Maybe that copy of Klimt's *Kiss* that used to be in the bathroom has been replaced by a poster of the Arsenal first XI; maybe the pot of organic yoghurt in the fridge is now hidden behind a six-pack of beer and a half-empty tin of beans. If this is so, we have found the seat of the problem: your boyfriend is giving up a nomadic existence and becoming sedentary.

Much though a debate between 'feminine good' and 'masculine bad' may fit the strange tastes of our society, foisting it onto our ancestors does not necessarily make good sense. The salient difference between lunar and solar cultures is not a matter of differing views on the necessity of personal hygiene or an interest in *Ally McBeal.* It is to do with the extent to which the participants in these cultures move around.

When your boyfriend was a nomad, Tracey, a solar calendar was no use to him whatsoever. As the Sun does not go through phases, the solar calendar is determined by the position of the Sun's rising on certain days of the year. But these points can be established and referred to only if the viewer is in the same place year after year. Your boyfriend might have noticed that on the Spring Equinox the sunrise occurs exactly at the top of the chimney that can be seen from Sharon's bathroom window; but if he is still a nomad and will not return to Sharon's bathroom, this information is useless to him. He can see the phase of the Moon from wherever he might be hanging his hat.

So your boy-friend's actions are less an attempt to impose a masculine life-style on you than a sign that he is settling down and becoming a sedentary culture. Your next question, of course, must be whether you want him to become sedentary right next to you.

Your caring, Neptunia

THE MOON AS MIND

One of the surprises on undertaking the traditional approach to natal astrology is to find how limited a role Mercury plays in the assessment of the native's mental capacity, and how great a role is that of the Moon. 'Whyever is that sensitive little bundle of emotions getting involved in things of the mind?' we may ask. So let us explore.

When judging the natal chart, we first determine the native's temperament. This is the basic 'how' of the native's existence: how he is in the world. The chart can be seen as an embroidery, in which the planets, each with its different coloured thread, weave the picture of the native's life. The temperament tells us whether this embroidery is on denim or silk or cotton, something of immense importance in our assessment of the garment.

Once we have established the temperament, we turn our attention to 'the wit and manner'. 'Wit,' of course, refers to mental capacity rather than the ability to make smart remarks. 'Manner' is in practice not dissimilar to temperament: it is another assessment of 'how', showing the manner in which the native behaves. But this is at a different level than temperament: manner can be acquired and refined; the scope within which we may alter our temperament is extremely small, for temperament manifests not only in character but also in bodily form: to change the one involves changing the other. No matter how long the sanguine type may spend in the gym, he will become only a sanguine type with some muscle; he will not turn into a choleric type. The word 'complexion', now used solely of bodily form, was once synonymous with temperament. One way of seeing the difference between manner and temperament is as motive and method: if we find that the native's manner is vicious and destructive, this is, as it were, his motive, his desire; the temperament will tell us how he goes about being vicious and destructive: the choleric temperament will lay waste with fire and sword; the sanguine temperament will be an intellectual iconoclast.

The assessment of the native's mental qualities is an important stage in the overall judgement, as it seems to be human nature that even the best-looking among us find flaws in our appearance that convince us we are ugly – the cellulite and sticky-out ears that no one else notices until we point them out – but even the densest of us are quite convinced they have the wisdom of Solomon. As the present writer is himself possessed of a strongly dignified Mercury it is tempting to accept that this is a sign of intellectual brilliance: but not so. For in our tradition, the idea of what the mind is and how it should be ordered is radically different from the ideas common today.

In the past, Mercury knew his place. It is not coincidence that he is also the planet of servants, for that is his role: Mercury, the reasoning faculty, is a servant, who needs to be strictly disciplined and can be trusted only to carry out minor day-to-day tasks, such as totting up the shopping-list to see if we can afford another pound of pears or programming the video-recorder. It is not his place to run the household, and even his advice is to be treated with the utmost caution.

It is easy to relate the Moon and Mercury as significators of the mind to the right brain/left brain split of modern popular psychology. This is much what Lilly means when he says that Mercury 'governs the rationall Soule and animal Spirits in the Braine, as the Moon doth the vegetative and strength of the Braine, more neer to the Senses.' This gives an approximation to their true function, but it is far from the whole story.

We must start with knowledge. What do you know?

- I know that the battle of Waterloo took place in 1815.
- No you don't. You accept what someone has told you about two virtually meaningless concepts.
- Well then, I know what I had for breakfast.
- No you don't. You trust that your memory is sound when recalling some sense impressions which may or may not have been correct.
- and so on...

What we *know* is what is revealed and accepted within the heart. The seat of mind throughout the world's traditions is in the heart (Sun) rather than the brain, which, like Mercury, is servant rather than master. So our knowledge is the Sun. It doesn't come out much nowadays.

If our Moon is functioning properly, it reflects (for such is its nature) the light of this knowledge, hence the desirability of finding Sun and Moon in harmonious aspect. Its proper function is sometimes referred to as 'intuition', although the common meaning of this word is a rather lower form of perception. Let us consider an example: the scientists would have us believe that one of our ancestors was wandering through the primeval jungle when he stumbled upon a chilli bush. Not having seen one before, he bit into a bright red pepper and thought 'Yum! I must put this on my dinner.' What our gourmet forbear really did, of course, was to contemplate the nature of the chilli pepper. The Sun of revealed knowledge was burning away brightly in his heart (much preferable to the fire of empirical knowledge singeing his innards), burning off the impurities of his mortality and so enabling him to see clearly. His Moon was well polished, and so reflected the light of the Sun out into the world and his perception of it (note the classical idea that vision is achieved by rays emanating from the eyes, not going into them as the scientists would have us believe. This is philosophically correct.). He was thus able to understand the nature of this fruit and for what it could be used, without the painful necessity of biting into it.

Centuries of neglect leave this lunar process of mind working usually in only the most trivial forms – as when we contemplate the nature of the meal and know which wine would suit it best. But even here, Mercury strives for mastery, and we are now more likely to learn by rote that a wine from this grape goes with a dish of that food. Thus it is that our idea of Moon/Mercury has deteriorated into the opposition of left-brain and right-brain.

We find many comparatively clear examples of Moon thinking in the literature of scientific discovery. It is a common phenomenon that the scientist's intense concentration on a problem will suddenly produce a non-rational apprehension of its solution. The scientist may not have the Sun of revealed knowledge burning in his heart, but it is as if the intensity of his application mirrors the process of nuclear fusion, creating a brief artificial sun by whose light he can understand. The accounts usually continue with him working out (Mercury) how to reach the answer that he now understands (Moon) to be correct. Perhaps the best-known example of Moon-mind in the scientific literature is Kepler finding his laws of planetary motion through the contemplation of the structure of the universe, as expressed in the perfect solids – though this was at a late stage in the decadence of science, and so the purity of even his understanding is sicklied o'er with the pale cast of thought.

In the traditional model of the ordered mind, Mercury has more to do with articulation than understanding: Mercury is the messenger, not the message – and we most surely do not give the messenger licence to make up the message for himself. Particularly if he has the dubious character of Mercury. No matter how strong he may be, he remains a shifty little so-and-so, and as such is to be treated with caution; if we doubt this, let us look at the Norse myths and the trouble caused by Loki (Mercury). Mercury is as utterly devoid of any sense of morality as his modern manifestation, the computer: he will process whatever is put into him. Reason can come up with a justification for absolutely anything, whether you are proving to yourself just why that last chocolate biscuit must be yours, or finding that the relative dimensions of somebody's skull give cause enough to murder him. It is a modern misconception that a strong Mercury and a sound Mercury are the same: if we, as astrologers, accept this we fall in with the scientists, who claim that Mercury can be trusted and that reason piled on reason makes a solid structure. This is only too evidently not so. Instead, we might remember that the Ragnarok, the end of the world, arrives when Loki/Mercury breaks free from the chains in which the other gods have wisely bound him and starts trying to run things: much what we see around us today. It was with words alone that Satan tempted Eve.

In our modern world, morality seems separate from mind: a man of the most dissolute morals may have a brilliant intellect. In the tradition, this is not possible. A mind is brilliant only if it is working correctly; if it is working correctly, the morals cannot be dissolute. The correctness against which the mind is measured is a higher concept: this itself is not a popular idea today. Better a dim servant who does what he is told than a brilliant one who thinks he is his own master; and Mercury, the reason, is our servant. It is the Sun and the Moon who are Lord and Lady of the House. So is it only fitting that astrology is ruled by Mercury, for if we could understand the phenomenon by our brightly polished Moon reflecting the knowledge of a burning inner Sun – and No, not by 'psychic powers' or any of the other dross that is trotted out to avoid making an effort – we should not need our charts and tables and other quaint gear, tools necessary precisely because we are not what we could be.

WHY IT GOES WRONG

There is an enduring illusion among students that their study of astrology is furthered by reading astrology books. In fact, apart from the perhaps half-dozen titles that make some substantive contribution to astrological knowledge, most books that repay the student's attention will not be found in the local branch of Stars-R-Us. Many, indeed, of the most worthwhile of these books make no mention of astrology from beginning to end. Yet they offer far more of value than the latest volume of much-the-same from the reprint houses or, of course, the latest work of sentimental fiction from any of the moderns.

The most illuminating astrological text that we have found recently is Norman Dixon's *On the Psychology of Military Incompetence*[3]. As the title suggests, it attempts to unravel the causes of the recurrent military propensity for blunder – without resorting to transits of the outer planets in order to do so. Dixon spent ten years as an officer in the Royal Engineers, turning to psychology, in which he has high qualification, after leaving the army. The resistance which the writing of this book met from certain sections of the military offers an immediate comparison with the world of astrology: suggesting that anyone might be capable of error clearly impugns their 'professional reputation' in writ-worthy fashion. The comparison between astrology and war is emphasised by the epigraph, from T.E. Lawrence: 'With 2,000 years of fighting behind us we have no excuse when fighting, for not fighting well.' Indeed; yet there can be few of us who do not look back on astrological troops whose lives we have squandered, and who are not, on our day, capable of misjudgements as gross as those on the Somme or Spion Kop. Theoretical knowledge and good intentions do not an astrologer make; nor does experience in no matter how great quantities give us the shield of infallibility.

So why do things go wrong? We may not follow the Freudian Dixon in locating the root of all evil in toilet-training, but the secondary causes he suggests ring as true in the astrologer's den as they do on the battlefield. 'The ideal senior commander,' he says, 'may be viewed as a device

[3] Jonathan Cape, London, 1976; reprinted Pimlico, London, 1994.

for receiving, processing and transmitting information in a way which will yield the maximum gain for the minimum cost. Whatever else he may be, he is part telephone exchange and part computer.' Much like an astrologer. 'On the basis of a vast conglomerate of facts (i.e. as drawn from the chart) coupled with his own long-term store of past experience and specialist knowledge, the senior commander makes decisions that, ideally, accord with the directives with which he has been programmed,' or are, in an astrological context, true.

Dixon gives two main reasons why this ideal is hard to realise. First is the number of incompatible roles that the commander/astrologer has to fill: 'These include 'heroic' leader, military manager and technocrat... politician, public relations man, father-figure and psychotherapist.' Second is the existence of 'noise in the system', interfering with the smooth flow of information that must be processed if a sound decision is to be reached. Noise is of two kinds: internal and external. For the military commander, an example of external noise would be the enemy bombardment beneath which he is sheltering; internal noise might be his regiment's proud tradition of never retreating, or the unfortunate fact that he is drunk. Any source of noise makes it that much more difficult to determine the best course of action.

Let us deal first with the nice causes: those for which we may plausibly blame others. The telephone-exchange part of the astrologer's nature faces inordinate difficulties with the human tendency to hear only what it wishes to hear. In the workshop, it is now our practice to tattoo all judgements onto the client's forehead, as any other means of communication is open to distortion. We have been criticised for getting predictions wrong which were clearly and verifiably right, and – in roughly equal measure – been praised for getting predictions right which were clearly and verifiably wrong. So making our own communication as clear as possible must be a priority: doing the astrology is not sufficient if we are unable to effectively communicate our result (although as even the apparently lucid '1-0 to Italy' is capable of misinterpretation, there are limits to our own responsibility for this). Or we can solve the problem by adopting the modern technique of never saying anything of any substance: the equivalent of keeping all our troops in barracks for fear of losing a man.

Our telephone exchange must be clear also about in-coming calls. If all our information tells us the enemy is in the east, it is reasonable that we do not expect an attack from the west – as long as we have ensured that our intelligence services are up to scratch. If the client chooses to tell us he was born on June 3rd when he was really born on May 4th (not an unknown occurrence!), there is not a great deal that we can do about it; but in many situations we can act to sieve the incoming information, always remembering that we too are prone to hear only what we wish or expect to hear. It is easy to leap to unwarranted assumptions about the client's situation or his attitude towards it. We must be aware that the client's words can act like triggers, setting off whole stories in our heads that have nothing whatever to do with the client's own experience. Even if we appear to be labouring points of painful obviousness, time spent interrogating the client to assure ourselves – as much as is ever possible – that we are answering his question and not one of our own is time well spent. The client is not ourself, and we are unlikely to exaggerate the possibilities of just how bizarre or extreme the differences can be.

We too play a variety of incompatible and largely unhelpful roles. The 'heroic leader' part which has heroically led so many soldiers to a needless death corresponds to the 'omniscient sage' who must be able to provide an accurate answer to every query. In William Lilly's army we are allowed to display the better part of valour without being left alone with a loaded revolver as a consequence. We are managers in a way that is no longer common in the military: the general does not have to pay the rent on the barracks; the necessity of our doing so is one of the loudest sources of noise that we encounter. We are technocrats, and must resist the temptation to play with our favourite toys of technique or software in circumstances where their use is unhelpful. We are our own public relations men – and the claims of our internal PR department can conflict with those of sound astrology ('Die? You? Never!') Most of all, there is but a small proportion of our clientele that does not to some extent attempt to ease us into the roles of father-figure or psychotherapist, even if we have the good sense not to seek out these roles for ourselves. Still more parts beckon us: Man of Mystery and Santa Claus are two of the more seductive ones. The conflicting demands of these roles are no more helpful to the astrologer than they are to the general; if we cannot shed them all, we do well at least to be aware of them: a large mirror is a useful tool. Thomas Merton was speaking of spiritual vocation, but his words are none the less relevant to us labourers at the forge: 'In any vocation at all we must distinguish the grace of the call itself and the preliminary image of ourselves which we spontaneously and almost unconsciously assume to represent the truth of our calling. Sooner or later this image must be destroyed and give place to the concrete reality of the vocation *as lived* in the actual mysterious plan of God, which necessarily contains many elements we could never have foreseen. (Our work involves) learning the fatuity and hollowness of this illusory image, which was nevertheless necessary from a human point of view and played a certain part in getting us into the desert.'

'I must have silence in my passages.'

Let us consider in detail the problem of noise, external sources first.

The commander might well suffer from unclear directives from above; so too the astrologer. What does the client want? Does he want to know what will happen, or does he want a suggested course of action, or does he want to know how he got into this mess? Or maybe he just wants validation. We may not choose to offer all these services, as we may choose not to send our troops over the top for insufficient reason; we do need to be aware of which battle we are fighting.

The astrologer is blessedly free from noise caused by enemy action. The inadequacy of his intelligence sources is down to him to rectify by sharpening his interview technique. There remains the possibility that his computer has been subverted by the enemy: our own has been brainwashed into believing that Miami is a town in the north of England, while Lancashire appears to share a border with the Ukraine. What goes into the computer is not necessarily meaningfully connected with what comes out, so this does need to be checked.

More significant in the eremitical stillness in which we labour is the internal noise, sources of which are manifold and often apparently unavoidable. The virtue of astrology is that it promises a 'noise-free' understanding of the situation, circumventing the illusions of hope, fear and desire that prevent the client understanding it for himself. He is not well served if all he has done is exchange his own noise for the astrologer's: a silent environment must be our ideal.

It is this end to which the Considerations before Judgement are directed. Some are concerned with external noise: the problem of inaccurate time-keeping is handled by the caution on judging charts with early or late degrees rising. If our timing is accurate, these cautions can be ignored. Some are concerned with internal noise, particularly those regarding the condition of the seventh house and its ruler (i.e. the astrologer). If we have a burning temperature, or have recently experienced some personal trauma, or have all our attention focussed on the man-eating spider crouching beside our chair, we are unlikely to give sound judgement; thus if the Lord of the seventh is afflicted, we see the astrologer below his best. (Let us deal here with a pervasive error: if you are casting a horary for your own question, you are *not* shown by the Lord of the seventh because you are also the astrologer. How many houses do you think you're entitled to?)

We will assume now that our astrologer is healthy, sober and untraumatised. The internal noise, however, is still at high volume. Shouting loudest is often the demon Self-Regard. His forms are manifold; his weapons those stored in the 7th/5th, 2nd and 10th houses. The astrologer is not a machine, but a human being attempting to act with a mechanical degree of dispassion. We may not often find the glamorous Fifi entering our consulting room and asking 'I had a dream that I must throw myself at the first astrologer I meet; is this dream true?' but we must be aware of the noise-producing qualities of a pretty face, as in any circumstance of the natural preference for being liked, usually as the glad bringer of good tidings. Our personal practice is to avoid seeing clients face to face, working instead by phone or post, for just this reason; but while this may muffle the volume, it certainly does not impose silence.

Then there are the tenth and second house temptations. The desire for reputation is pernicious; the desire to give right judgement is a hindrance. Our focus must be on the chart, not on its consequences for ourselves. The dangers may not be immediately obvious: but consider the general whose desire is to be seen as the Heroic Victor or the Winner of All Battles, and think of the effects of these desires on his decisions.

William Lilly stormed off from his first teacher when he found him altering his advice to a client because 'had he not so judged to please the woman, she would have given him nothing'. It is warming to think that as we would not so grossly betray our art, we are above the clamour of money. Unless our gentle reader is significantly more saintly than ourselves, however, this is unlikely to be true. We need not grasp at money for it to be an issue: the awareness that we would like our client to come back again – and the awareness that our client doubtless knows this too – is enough. As an example, whenever we give judgement that 'No, this is not a suitable person to employ,' we hear a small voice suggesting that the client will think we judge thus only to ensure ourselves of another question about the next candidate. Any such voice is a distraction from true judgement.

Ann Geneva, from whom as an academic we do not expect any understanding of practical astrology, and her followers, some of whom ought to know better, quote Lilly's annotation on a chart he had evidently got wrong: 'God Almighty bless me from the punishment due to an auld dissembler. Money, Money.'[4] Displaying a remarkable lack of any introspective capacity, they see the anguished Lilly in sackcloth and ashes after having prostituted judgement. Quite clearly, what we have is Lilly lamenting that his judgement has been clouded by the financial noise inevitable with a paying client. When applying for a William Hill Award[5], the very fact of the financial involvement distorts judgement; we have found our ability to predict accurately improving no end since we stopped applying for awards ourselves; yet it is obvious that the distortions of judgement in these circumstances are detrimental to one's own interest: judgement wrong – money lost.

Dixon identifies an anally-inspired fixation on bull as a major source of noise: an obsession with doing what 'ought' to be done at the expense of what make sense. He makes the point that there is often a high turnover of senior commanders at the start of a war, as after a long period of peace these senior positions are perforce occupied by those whose only knowledge of the skills these ranks demand has been gained on the parade-ground. They often fall short in practice, and those who talk a good battle are replaced by those who can fight one. We might speculate that the appreciation of discipline and order that leads one man into the higher echelons of the military is not very different from the appreciation of discipline and order that leads another to study the traditional branches of astrology; as such, the astrologer faces the same danger of placing too great an emphasis on the letter, of denying his own perceptions too much, of passing the buck upwards (in his case to Lilly or Bonatus) and of opting out when faced with the novel possibility of employing his brain. It is notable that the powerful saturnine influences common in those who study traditional astrology are also prevalent in those whose work we study (see Lilly's nativity as an example), suggesting that it is the order rather than the antiquity of the knowledge that contains its appeal. But if the gifts Saturn offers are not leavened with

[4] Ann Geneva, *Astrology and the Seventeenth Century Mind*, Manchester University Press, 1995.

[5] [Awards for accurate sporting predictions. Details of how to apply can be found in most issues of *The Astrologer's Apprentice*. William Hill – for the benefit of foreign readers – is a leading book-maker.]

sound common sense, they become strait-jackets within whose grasp we cannot move. This is one form of noise by which the moderns, with their 'make it up as we go along' philosophy are less troubled; working within a tradition we need to be constantly aware of it: there is a fine line between discipline and stupidity. The generals who had learned to form their troops into squares to face cavalry found them massacred by machine guns; too often we find a slavish adherence to the letter of even the best of our authorities resulting in statements that are palpably daft.

We shall skim over some of the other recurrent features of military incompetence that Dixon identifies. We have 'the failure to observe one of the first principles of war – economy of force' manifesting in the astrological world as a tendency to waffle interminably, either to obscure whatever definition one's judgement might ever have had or to delay the dreadful moment when one has to say something concrete. This tendency is notable in many computer programs, even those that claim to help the horary astrologer, supplying him with so much irrelevant information that the client has turned safely senile long before that risky moment of giving judgement has ever arrived.

'A tendency to reject or ignore information which is unpalatable or which conflicts with preconceptions.' This is most evident in judging one's own charts, where the most tenuous of favourable testimonies assumes overwhelming significance, while the massed ranks of negative indicators are completely ignored. But it can occur in any chart, usually when we are building a happy little story, testimony by testimony, and then suddenly notice an exact opposition from Saturn to our main significator, an aspect which had somehow avoided our eye till then. A moment of battlefield crisis: do we pretend we haven't noticed the Russian guns and let our cavalry ride up the valley, or do we have the courage to change our orders to accommodate this new information? Dixon quotes Festinger: 'Once the decision has been made and the person is committed to a given course of action, the psychological situation changes decisively. There is less emphasis on objectivity and there is more partiality and bias in the way in which the person views and evaluates the alternatives.' This decision can be made at any time – sometimes, sadly, before the chart is even cast, based solely on the astrologer's personal prejudices and assumptions. The earlier in the judging process it is made, the worse the judgement is likely to be, as it will consist merely of distorting or ignoring testimonies to make them fit.

There is the 'failure to exploit a situation gained' and a tendency to 'pull punches rather than press home an attack', as if it is not quite the decent thing to strive to one's utmost, or to apply the *coup de grace* to a chart that we have managed to wound. Hedging and demurral may perhaps be virtues in the Diplomatic Corps, but not in either war or astrology. It is a familiar situation that the mobile units at the top of one's mind have achieved a break-through in a chart, but judgement is lost through the reluctance to commit the whole army to follow them up. If the mobile troops have achieved a break-through it is because a break-through can be made; we suggest that the astrologer who fails to drive home his advantage should be shot at dawn *pour encourager les autres.*

'Failure to make use of surprise' is an interesting astrological concept, suggesting many a strange consulting-room scenario. 'A belief in mystical forces' is rather more common. How often have we heard the claim that 'I'm an intuitive astrologer'? 'I'm an intuitive engineer – let me build this bridge.' Intuition is fashionable; one is not cool without it. It has little if any place in astrology, being almost invariably either an excuse for laziness and lack of knowledge or a synonym for prejudice and immovable preconception. The client's next-door neighbour has intuitions; from an astrologer he expects the truth.

'A suppression or distortion of news from the front' hardly needs comment. I, of course, have never made a mistake; but those astrologers who have are reluctant to admit it, even, often, to themselves. This relates to 'an undue readiness to find scapegoats' – every other astrologer being incompetent – and 'a distinctly paranoid element in the way some senior commanders have reacted to even the faintest breath of criticism; to the vaguest and most tactful suspicion of a raised eyebrow or cleared throat.' As it has now become unwise to mention any living astrologer without suggesting that he be immediately awarded the Nobel Peace Prize, it does seem that the military is not the only sphere in which the reaction to criticism is 'distinctly paranoid'.

There is far more of value in Dixon's book than we can begin to address here. It does, indeed, perform for astrological practice one of the more valuable functions of astrology on daily life: it holds a mirror to it, through which we can see it from an unfamiliar angle and grow in understanding. We strongly recommend anyone with any interest in practical astrology to set aside Spurious Superfluous's *Treatise on Nit-Picking, Part IV* and reassess their practice in this light; for once the basic methods are known, it is by refining our hearts, rather than piling technique upon technique, that we shall achieve mastery of our craft. Dixon enumerates many of the snares into which we are prone to fall, and from which we may escape only by making a final decision as to which of the two masters we wish to serve. William Lilly's advice *To the Student in Astrology* is more than just conventional piety, but a sound practical lesson in astrology, by following which we may become deaf to both the internal and external noises which cloud judgement. As Lilly tells us, 'the more holy thou art, and more neer to God, the purer Judgment thou shalt give.'

5

Advanced Techniques

THE ASSESSMENT OF TEMPERAMENT

The calculation of temperament is the vital first step in analysis of the nativity. Skipping this stage leaves us ignorant of the nature of the beast with which we are dealing. No matter what the other indications in the chart may tell us about the nature, omitting this initial step leaves us ignorant of whether the nature in question is that of a tiger, a gibbon, a bull or a giant sloth. This is a distinction of no small significance. Most of what we read from the chart, and certainly all that we read from it on the psychological and physical levels, must be read in the light of this initial distinction (if we decide that our native is a bad-tempered tiger we have more cause to worry than if he is a bad-tempered sloth!). Yet our ability to perform this calculation is hampered by the texts – and by the modern obsession with performing arithmetic at every conceivable opportunity.

Lilly's worked example of the method in *Christian Astrology* is badly misleading. Even Lilly admits that he couldn't get the answer right, as having carefully worked out the temperament to find that hot and dry (choler) predominate, he abandons this conclusion and announces that the native is sanguine melancholic, conjuring up unlikely testimonies to cobble together the answer that he knew to be true.

The error is in the way the various testimonies are weighted. Among the human flotsam that has been washed up in our yard over the years and to which our Master, being a kindly soul, affords a home, is a tattered old sea-dog called Harry Stottel. He claims to have been helmsman of the good ship *Reason,* which floundered with disastrous loss of life after a collision with a pirate raider, *The Enlightenment,* sailing at full speed in the opposite direction, at some time towards the end of the Eighteenth Century. (Incredible as it may be after so devastating a collision, those who have seen *The Enlightenment* recently say that it bears not the slightest sign of ever having been in contact with the *Reason.*) Harry now slumbers his twilight years away before the refectory fire, except when awakened by the stable-lads lighting tapers placed between his sleeping toes.

On one of these brief excursions into consciousness, he muttered something about nature and accident that clarifies our present task. "Let us take a noun – 'man', for example. We may qualify this noun by the addition of adjectives. But no matter how many adjectives we might add, the noun remains the same. We may say the man is womanly, but no matter how strongly this description may qualify his nature, he remains a man who happens to be effeminate; he does not become a woman." At this point one of the stable lads giggled something about 'California' – whatever that might be – but he was promptly carried off and douched under the yard pump to restore him to his senses.

Lilly's error is to weight adjectives as strongly as nouns. For example: suppose one of our testimonies is a hot, dry planet. This is our noun. It is in a cold, moist sign. This is our adjective. So our planet will have trouble being as hot and dry as it wishes to be: the cold and moisture of the sign diminish the heat and dryness of the planet. But they cannot completely obliterate this heat and dryness, nor can they change them into cold and moisture. *The adjectives qualify the noun; they do not alter its fundamental nature.*

The Method

There are four factors to be judged in determining the complexion or temperament:

* The Ascendant and its ruler
* The Moon
* The Sun
* The Lord of the Geniture

The qualities of the planets are:

♄	oriental	cold & moist
	occidental	dry
♃	oriental	hot & moist
	occidental	moist
♂	oriental	hot & dry
	occidental	dry
♀	oriental	hot & moist
	occidental	moist
☿	oriental	hot
	occidental	dry
☽	first quarter: hot & moist	second quarter: hot & dry
	third quarter: cold & dry	last quarter: cold & moist
☉	Spring (♈♉♊): hot & moist	Summer (♋♌♍): hot & dry
	Autumn (♎♏♐): cold & dry	Winter (♑♒♓): cold & moist
☊	as Jupiter	☋ as Saturn

Note that the seasons of the Sun repeat the phases of the Moon. Occidentality removes temperature from the planets' gift: if the planet precedes the Sun it has the chance to warm or cool as it pleases; if it rises into the sunlit sky, it can do neither.

The first and most important rule in assessing temperament is to abandon all thoughts of nicety of detail. Grab the Virgo in your being and find it something else to do, or you will be old before your time. This is rough-work: we do not use tools built for micro-surgery. Don't worry: you won't go wrong!

For our worked example we shall use the nativity of Roy Orbison.

First point: The Ascendant and its ruler.

Asc is Virgo: cold and dry. This must be qualified by any planets in the first house. There is none. In practice, we can safely ignore anything that is in the first but in a different sign to the Ascendant, and anything that is more than, say, 4-5 degrees from the Ascendant.

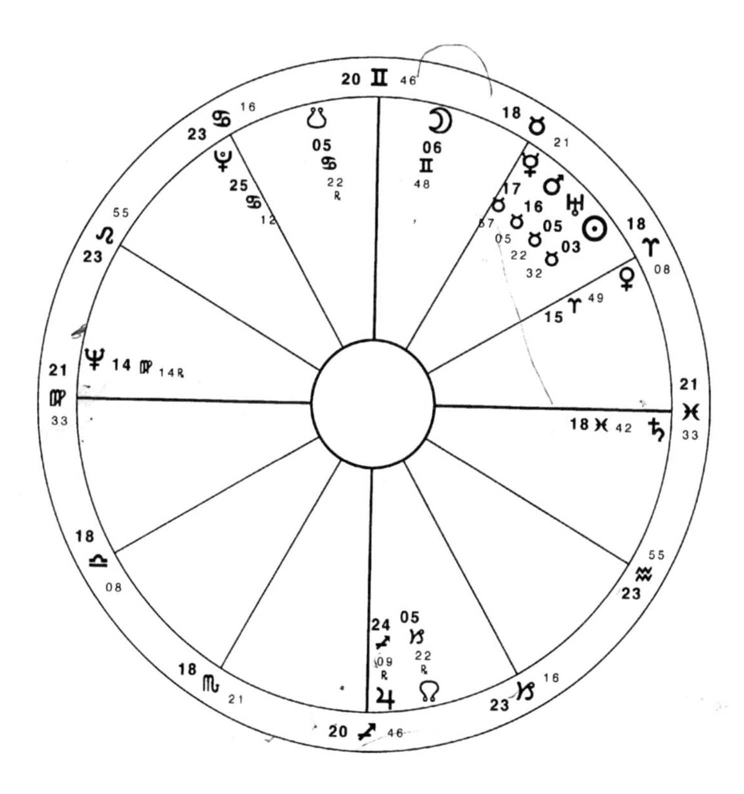

Chart 9. Roy Orbison. April 23rd 1936. 3.50 pm CST. Vernon, Texas.

Now qualify by planets in close aspect to the Ascendant. 3-4 degrees is our limit here.

Saturn opposes. Saturn oriental: cold & moist. In a cold, moist sign: very cold and very moist. This will add to Virgo's coldness and diminish its dryness. If Virgo is CD, we now have C+ D-.

Those of you who have not followed our advice by finding other occupation for your Virgo nature will want to qualify Saturn's influence further by factoring in its own aspects (sextile Mars & Mercury). Don't: we'll be here for ever, with no tangible increase in accuracy.

Jupiter squares. Jupiter oriental: hot & moist. In a hot, dry sign: very hot and only slightly moist. This will reduce Virgo's coldness and further diminish its dryness. So we now have C D - -.

NB: in nativities we must pay attention to aspects *in mundo* as well as *in caelo*. That is, a planet on the MC or IC squares the Asc regardless of whether the angle between them is 90 degrees. So if Jupiter were on the IC and the IC were at 5 Sagittarius, Jupiter would still be square the Ascendant.

Mercury trines. At just under 4 degrees, this is as distant as we need consider. We would not usually bother with Mars here (too distant), but as he is keeping company with Mercury we must involve him too. Mercury occidental: dry. In a cold, dry sign: cold and very dry. Mars occidental: dry. In a cold, dry sign: cold and very dry. So the pair cools and dries the Asc. So our C D - - now becomes C+ D-.

We would also include antiscia, allowing a maximum orb of around 1 degree, but there is none.

Now for the Ascendant ruler. Back to Mercury again: cold and very dry. Cooled and dried further by the conjunction with Mars. So C+ D++

Sextile Saturn. As we have seen, this is very cold and very moist. So we now have C++ D+

So the Ascendant and its ruler are extremely cold and very dry. So far, our native is melancholic. C+++ D++

Second point: The Moon.

NB: when we consider the lights as part of our basic strategy of assessment we do not count the Moon as being cold and moist, or the Sun as being hot and dry, as these are givens in every nature. But – very importantly – when we bump into them along the way we must. So if, for example, the Ascendant ruler were the Sun (this not being a given in the nature), we would consider it as being hot and dry, qualified by its season and by its sign.

The Moon is in its first quarter: hot and moist. In a hot, moist sign: very hot and very moist. No aspects. So H+ M+. For one famous for *Crying*, we had to find some moisture somewhere, and here it is!

Third point: The Sun.

It is in a Spring sign, so hot & moist, qualified by the nature of the sign itself, which is cold & dry. So slightly hot and slightly moist. H- M-. (Had it been in Gemini, for instance, it would have been even hotter and even moister.) No aspects.

Fourth point: The Lord of the Geniture.

The Lord of the Geniture is the planet with most essential dignities, with the rider that an essentially weaker planet with accidental strength may be preferred to an essentially stronger planet in a difficult position (so we would choose a planet in its own triplicity that is on the MC over a planet in its own sign that is in the twelfth house).

There is no doubt here: Jupiter. As we have seen, it is very hot and slightly moist. No aspects. Had the Moon been where Saturn is here, we might just have considered that, as the Moon moves so fast and is applying, while Jupiter retrogrades to meet it. So H+ M-.

In Total.

The four points have produced:

* The Ascendant and its ruler	C+++ D++
* The Moon	H+ M+
* The Sun	H- M-
* The Lord of the Geniture	H+ M-

We have three Hots and three Moists, all modified for better or for worse. These both outweigh the one Cold and the one Dry, no matter how strongly these are able to manifest. Sanguine (hot & moist) wins. So the temperament is predominantly sanguine. There is a powerful melancholic (cold & dry) streak, however, and this must be borne in mind. Certain authorities would resolve all these distinctions into one coherent whole, but this sacrifices finesse: the fact that there is a dissenting voice in here is important, and considering the place from which that voice comes will tell us much about the nature. Our four points of judgment are not just different slices from the same cake, but are on different levels of the being. Suppose, for instance, that the first three points gave phlegm (cold and moist) while the Lord of the Geniture were melancholic: we might have evidence of someone with the potential to struggle against his powerful desire nature (phlegm) by striving to develop his contemplative faculties (melancholy).

In Orbison's case – briefly and broadly – we have someone with the potential to rise above what threatens to be a destructive melancholy through the rational (in the strict sense) faculties. How well and in what manner these rational faculties will function is a question beyond the scope of the consideration of temperament; this we can discover as we look deeper into the chart.

Each of the temperaments has its associated planets, the condition of which will tell us much about how that temperament will function. Saturn, being itself cold and dry, is the planet of melancholy. Mercury (cold and dry) can also lend a hand. A melancholic nature with Saturn strong in the cold, dry sign of Capricorn, well placed in the chart, might be a great contemplative. A melancholic nature with Saturn in fall in Aries, opposing the Ascendant, might well forget the canon 'gainst self-slaughter. Orbison's Saturn is peregrine and does oppose the Ascendant, hence the difficulties. Things could be a lot worse, but this testimony is entirely congruent with someone who became famous singing about how miserable other people (Saturn on 7th) made him.

The sanguine temperament needs Jupiter (hot and moist) or Mercury (by natural association) working well. Here, Orbison is blessed: Jupiter in its own sign and angular; so there is good cause to think – at this early stage in the judgment – that he will be able to rise above the destructive melancholy. Without such buoyancy to lift the melancholia the tragedies of his personal life might well have overwhelmed him. Even the Jupiter is retrograde, however: he was noted not for his rendition of *Zippedy Doo Dah!* but for a collection of mournful ballads.

We trust that even this brief look at the temperamental balance will make clear how invaluable is this knowledge for accurate judgement of the character.

SEXING THE CAT

As is well-known, the approaches of modern and traditional medicine are quite different. In terms that are not so very over-simplified, modern medicine aims to label the particular set of presenting symptoms as specifically as it can, and then suppress them. Traditional medicine, which has what seems to the modern ear a careless lack of concern for specific labelling, aims to restore the natural balance the failure of which is causing the presenting symptoms. To the modern doctor – with, again, only a slight over-simplification – the symptoms are the problem that needs to be resolved; to the traditional physician the symptoms are a warning that something has fallen out of balance. From the traditional perspective, the modern habit of treating the advent of such a warning by swallowing an anodyne is no different from shooting the messenger before he can tell his message. With the warnings so widely ignored, it is no surprise that the constitutional imbalance becomes ever worse, until it manifests in a chronic and incurable ailment.

Much that masquerades as holistic medicine by paying lip service to the above is in reality anything but. Treatment does not become any more holistic or any less symptomatic merely because you buy your potions from a health-food shop instead of a pharmacist. A good proportion of so-called holistic medicine concerns itself with the illness rather than the patient in exactly the same way as does its conventional brother.

Moving now from diagnosis to prescription, it is worth correcting the common belief that 'you can't do any harm with herbal remedies'. If you couldn't do any harm with them, it is unlikely you could do much good with them. Traditional medicine regards even herbal prescription as a drastic intervention that is best avoided if at all possible. If the native is looking after himself at all well, the condition should not have become so serious that it needs treatment with herbs.

The native is usually, of course, not looking after himself at all, and will do his utmost to continue with the same unsatisfactory habits that have led to the symptoms of which he complains. Dietary modification, then, which is at once the most effective and the most benign of treatments, does usually need the short, sharp shock of herbal support. In the workshop we open a vein or two – a key practice in traditional medicine – but this treatment seems no longer to be acceptable among even dedicated 'holists'.

Neither of the cases in these example charts called for prescription of any kind; both charts, however, illustrate some of the basic principles of astrological diagnosis.

The querent's baby had reacted badly to medication and no longer had either a sucking reflex or any interest in solid food. She was currently being fed by tube. Doctors did not regard her life as endangered. The querent wanted to know when she would develop an appetite and when she would be back to normal.

The baby is shown by the fifth house. Caput Algol on its cusp is just one of many unfortunate indications. The Moon has little light and is losing what little it has; the Sun is in fall; the Moon is heading first to square retrograde Saturn and then into the eighth house, where the

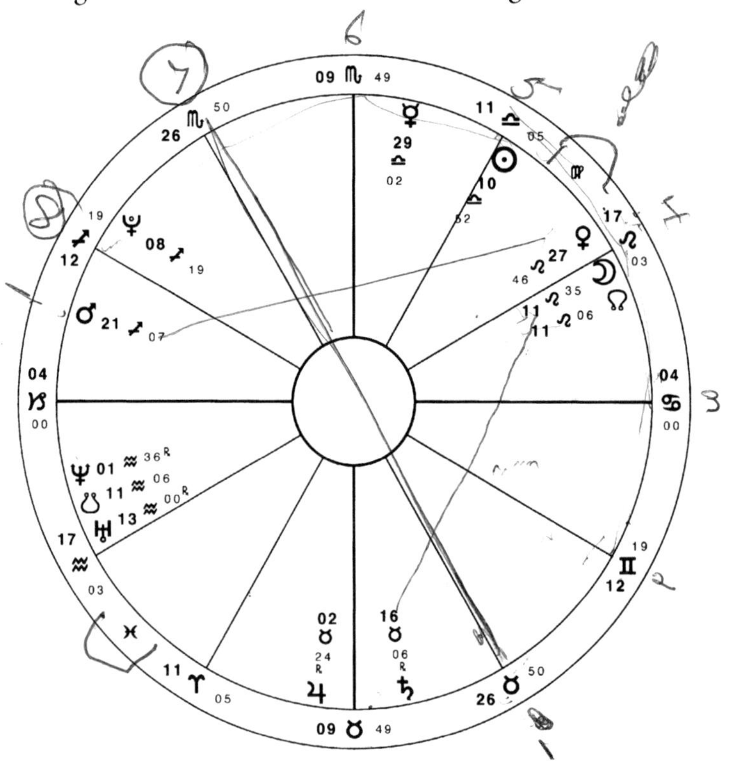

Chart 10. When will my baby eat? October 4th 1999. 2.39 pm BST. London.

baby's significator already stands. If death were decided by a democratic vote within the chart, the child would have little hope of survival. But it is not, and there are few cards so powerful that a mutual reception between the main significator and the ruler of the eighth house will not trump them. Such is the case here. As Lilly suggests, there will be recovery after despair.

Venus, the baby's significator, is cold and moist. It is in a hot, dry sign. This uncongeniality of placement shows that she is indeed ill. Peregrine, she has little strength. The next thing that Venus does is to enter Virgo. This is cold and dry: not perfect, but a marked improvement on Leo. In Virgo she will have dignity by triplicity, but she will also be in her fall. This too is far from perfect, but it does at least show something happening: this is probably as much as can be hoped for at this stage. The main significance of entry into Virgo, however, is that Venus is now in the sign and exaltation of Mercury. That is, the baby wants – and wants very strongly – whatever Mercury signifies. Mercury rules the turned second house. The second is the house of food. Entry into Virgo must show her recovering her appetite.

Venus must travel just over two degrees before entering Virgo. A reasonable time-scale for the question would be days, weeks or months. Months, the longest option, would be shown by a fixed sign in an angular house. Venus is succedent, so this must be quicker than that. So two weeks.

As Venus enters Virgo, a cold sign, we see the baby start to recover. The promised land of Libra, Venus' own sign, becomes visible before her. That must represent full recovery. Passage right across a sign indicates one or other of the basic time-periods, either a year or a month, the context making it clear which is the more likely. Here, it must be a month. So she will start eating after a fortnight and be fully recovered after another month. And so it proved.

What sex is the cat?

The querent had found a stray kitten wandering outside her flat. She had put down some food for it, but as the kitten was eating, the local boss cat arrived and chased it off. The question was: "Is the kitten all right? Will I see it again?"

The querent's concern for the cat is well shown by her significator, Mercury, placed on the cusp of the sixth house (small animals). Venus

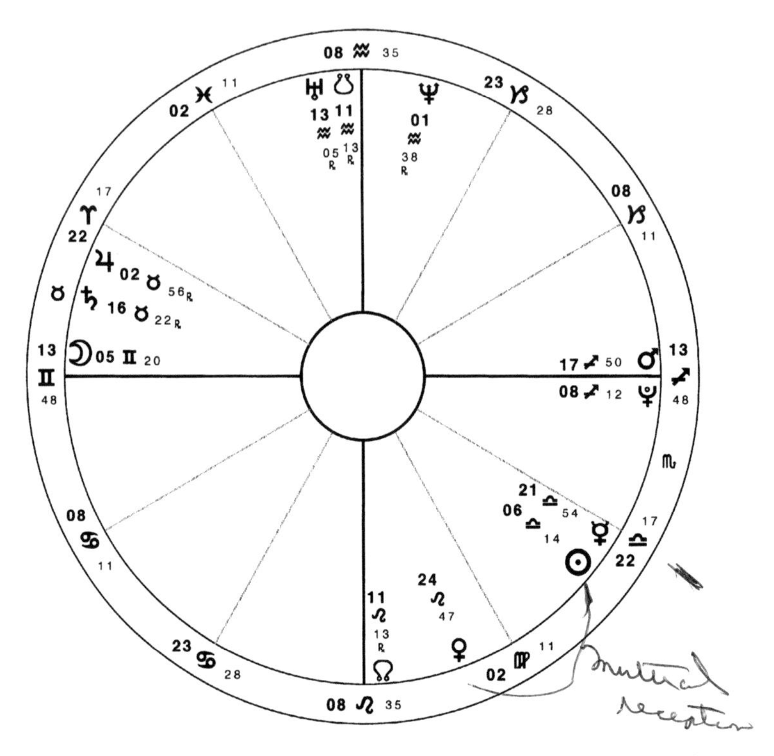

Chart 11. Fate of puss? September 29th 1999. 9.12 pm BST. London.

(the cat's significator) is peregrine in the fourth, showing the cat wandering about (peregrine) near the querent's home (fourth house). Spica, the most beneficent star in the sky, on the sixth cusp, an applying antiscial conjunction of Venus and Jupiter, and an absence of negative testimony are good indications that she will come to no harm. Mercury applies to sextile Venus. Mercury is in a cardinal sign and cadent house, suggesting our shortest feasible time-scale. The querent will find the cat in a matter of hours; and so she did.

Work done, we can now turn to play. The querent had wondered what sex the kitten might be. The only description that she had given was that it had something wrong with its left eye. The chart uses this point with beautiful simplicity to reveal the cat's gender.

The cat is shown by Venus. As in the chart about the baby, Venus is in Leo, an uncongenial sign, so we know that there is something wrong with the cat. We were not asked to provide a diagnosis of the baby's ailment; this chart gives a glimpse of the standard method.

The ruler of the sixth house usually shows the location of the problem rather than the diagnosis. This, rather, follows naturally from our first step by which we determined whether or not the patient is actually ill. In this example, the placement of the cat's significator in an uncongenial sign has confirmed that the cat is ill. That is, the cat is ill as shown by its planet being in Leo; so its being in Leo is the problem. The Sun, ruler of Leo, rules the eyes and is debilitated by being in the sign of its fall. The chart confirms what we have been told: the cat has something wrong with its eyes. The mutual reception between the Sun and Venus suggests that the problem is not serious.

Specifically, we know from the querent that the cat has something wrong with its left eye. The problem is shown by the debilitated Sun. The Sun rules the eyes in general, but in particular the right eye in a male and the left in a female. With its left eye affected, shown by the Sun, the cat must be female. And so it proved. Try posing this riddle to a vet: "This cat has something wrong with its left eye; what sex is it?"!

THE SURGERY

For all that horary allows us to peep behind the curtain that obscures the future from our curious gaze, it cannot be denied that in most cases this is a pointless operation. It does allow the astrologer to hone his technique; but for the astrologee what happens will happen, and the value of knowing about it in advance is – whatever dubious benefits the texts may rehearse in its favour – usually none.

The greatest qualifiers of whatever utility prediction might have are the fallibility of the artist and the client's awareness of this. Ars is ever longa, vita brevis, and application more brevis still; so even those predictions which might, the apologists claim, find a value in dispelling unwonted fears of disaster have this power taken from them by the client's knowledge that no artist is infallible. Fear is dispelled only insofar as the client maintains the illusion that he is.

So while horary is seen primarily as a tool for prediction, its greater value lies not in its ability to forecast the outcome of any situation, but in its ability to provide a clear and succinct analysis of that situation. We suspect that this consideration may, among other pressing causes, have had much to do with William Lilly's increasing concentration on medical astrology, for the medical chart is where this analytic ability is seen at its clearest. And the limitations on prediction – for if, suppose, we read from the chart that the patient will die, are we not then to treat him in the hope that he may not?

The recent passage of Saturn through the latter part of Taurus brought the sick and decrepit of the kingdom to the workshop gate, thrusting aside even the broken-hearted maidens who make up the bulk of our clientele. As is usual, it was the final of the planet's three passes over this section of the zodiac that brought its strongest manifestations: the first direct and the retrograde passage soften us up; if we haven't taken heed by then, the second direct passage hits us. Thus also with transits.

As we might expect, the stars inspire questions that match their current pattern. Once Saturn had moved into Gemini, the legions of the ill faded like the dew, to be replaced by Baron Hardup and his crew posing questions on financial and property matters: the knot of conjunctions and oppositions between Mars, Mercury, Jupiter and the Sun presumably being so tangled and troublesome that these questions alone would fit it. Saturn however, brought us the sick.

'Why?' we might wonder. For Saturn is quite at home in the latter part of Taurus. It has dignity by face through all the final ten degrees, and by term through half of them. It is in a sign of its own nature: a cold, dry planet in a cold, dry sign. Our first question to any medical chart is 'Is the main significator in a sign of its own nature?'; if it is not, we have a clear picture of the patient out of sorts. We should, then, in principle have a fairly trouble-free Saturn when it is placed here.

Far from it. Things must ever be judged in accordance with their nature. Take the matter of speed: being swift in motion is an accidental dignity, making a planet stronger. We must be cautious with this when it comes to Saturn: moving fast is against its nature. For Saturn, being swift in motion is likely to render it unstable. So here: for all that there is much to be said for Saturn in this position, it is excessive. After all that time plodding through the fixed, earthy sign of Taurus, the stuckness that is Saturn became far too stuck for anybody's benefit.

In medical terms, Saturn embodies the retentive principle. This is necessary for our well-being. 'Retention in digestion,' Saunders tells us, for instance, 'Is to detain the meat in the proper place of digestion, till it be thoroughly digested to the conserving and strengthening of nature.' After digestion, retention is performed by all the various bodily members, who retain the nourishment so that they can make use of it. What enables them to do this is the action of natural melancholy in the body. The obvious manifestation of an excess of the retentive faculty is constipation; that of an excess of the expulsive faculty is diarrhoea. The same causes have similar manifestations throughout the being, and on various levels. What it is now common to describe in Freudian terms as an 'anal retentive', for instance, has, in traditional terms, an excess of the retentive faculty manifesting on an mental/emotional level (and, given half a chance, on the physical as well). In the traditional model, it is the excess or weakness of this faculty that is at the root of most mental and psychic disorder.

Saunders tells us that Saturn in the last twelve degrees of Taurus brings 'visions and fantasies, melancholic passions, solitariness, heaviness and sadness', along with various physical manifestations. In this third passage across these degrees, especially when passing Caput Algol, the mental side of this seemed most commonly in evidence. This question is an example. The querent's sister's boyfriend was suffering from a mental condition of some duration. The question did not relate directly to the underlying condition, but to the drugs he was taking to treat it. These were controlling the condition satisfactorily, but were having disorienting side effects.

The texts are confusing over our way in to such a chart, when the question is asked by a third party. In some places we are advised always to give the first house to the patient; in others, to give the first to the querent and to assign the patient to whichever house he would have in any other horary. This is correct practice: if the chart is for a horary question, there is no reason to give the patient the first house unless he is asking the question himself. The confusion comes from the texts habitually treating decumbiture charts and horaries in the same breath: in a decumbiture, the time of which is taken from the event of the patient

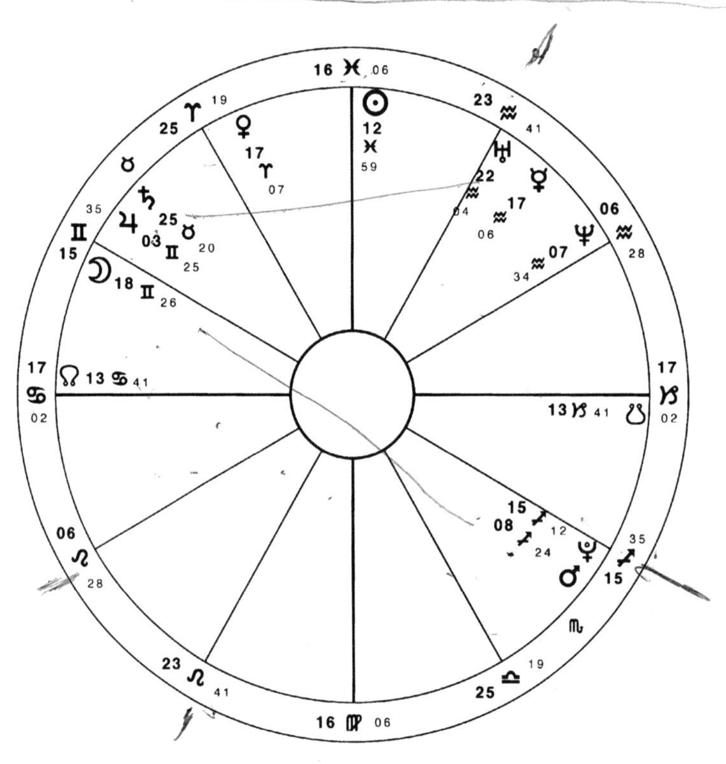

Chart 12. Side effects? March 3rd 2001. 12.24 pm GMT. London.

becoming 'so ill, or so extremely oppressed that he was enforced to take his bed', the patient does always have the Ascendant. Event charts are event charts and horaries are horaries.

Here, then, the patient is the querent's sister's boyfriend: the seventh house (boyfriend) from the third (sister) from the first (the querent), bringing us to the ninth. So his significator is Saturn, the ruler of this house. Outer planet fans will delightedly descend upon Uranus, placed just on his cusp, and find therein all the judgment; we have not found Uranus to be significantly involved in such matters.

The position of the main significator, with the manifestations of the excess of retention, as above, exacerbated by the placement on Algol, describes the patient's condition well enough. It is not sound in medical charts to rush to the ruler of the sixth as significator of the illness. We are better to head for the planet that is causing our significator problems. What is causing Saturn's problems here? Its position in Taurus. This is confirmed by the position of its dispositor: Venus is in Aries, debilitated itself and, more importantly, receiving Saturn into its fall.

It is always worth looking at the Moon's most recent aspect. Lilly gives a detailed and generally reliable table showing the medical effects of the Moon's squares and oppositions from Saturn, Mars and Mercury. In fact, we need not confine ourselves to squares and oppositions: any aspect will produce much the same results. It is to be regretted that Lilly limited himself to these three planets, as the others, especially – but not only – when debilitated are quite as capable of causing ills. For the Moon in Gemini separating from Mercury, the disease is 'occasioned by weariness of the mind, and over-burdening it with the multiplicity of affairs'. Combining this with the indications of excessive retention we see that the patient has a kind of psychic constipation: experience is not being flushed through the mental system as it should.

But the illness itself is not our concern here: it is the side effects of the treatment. To treat this illness, occasioned of an excess of cold and dry, we would apply heat and moisture, working, Saunders suggests, with the energies of Jupiter (hot and moist) in any of the air signs (hot and moist). The treatment is shown by the tenth house. Turning the chart, this brings us to the radical sixth. The treatment that he is receiving, then, is signified by Jupiter in Gemini. This is most interesting, as it suggests that the modern doctors are applying, in different form, exactly the sort of balancing treatment that the traditional model would suggest.

Jupiter is in its detriment in Gemini, however, so the efficacy of this treatment must be limited. The side effects of which he complained were stiffness and other difficulties with walking. These are entirely congruent with the typical manifestations of excessive melancholy: 'cankers, gouts, and stiffness of the limbs and sinews'. This shows that the side effects are not generated by the drugs. The drugs are channelling the symptoms of the illness, suppressing some while allowing others – apparently, to the modern eye, unconnected – to remain. As the overplus of melancholy must out somehow, the symptoms that remain become exaggerated.

From the traditional standpoint, we see that while the principle of the modern treatment is correct, it is not being applied at the correct level. That is, it is meeting cold and dry symptoms with hot and moist remedy; but symptoms will ever multiply as the underlying cause seeks a route by which to escape. The traditional physician would have treated the cold and dry cause with hot and moist remedy.

A different excess is shown in the second chart. It is notable that this question, although again relating to a long-standing condition, had to wait until Saturn had finally moved out of Taurus before it could be asked.

The querent, who was asking about his own condition, is signified by Mars. Is the patient ill? Hot dry planet in hot dry sign: in theory, this is OK. But the planet is retrograde, suggesting that something is amiss. Mars has only just turned retrograde, and so is almost motionless against the sky. This provides, in another way, a similar picture of stuckness to that which we have considered above. The bodily principles should flow smoothly; stagnation is never good for them. Mars, stagnant in a hot/dry sign, gives us a picture of lots of choler with no channel for release – much like a crowd of people trying to squeeze through a narrow gap: the natural flow is interrupted as none of them is able to pass.

Our first suspect for significator of the illness is the dispositor of the main significator, in this case Jupiter. Jupiter is confirmed as the culprit by the Moon, which is separating from aspect to it. This is an illustration of the need, mentioned above, to consider all the planets, not only Mars, Saturn and Mercury, from which the Moon separates.

Jupiter is a hot moist planet in a hot moist sign: again we have an image of excess. 'Hot and moist' leads us to the blood: there is, as it were, too much blood. Jupiter signifies the expulsive faculty, and with such an indication we would expect symptoms such as nose-bleeds as the body

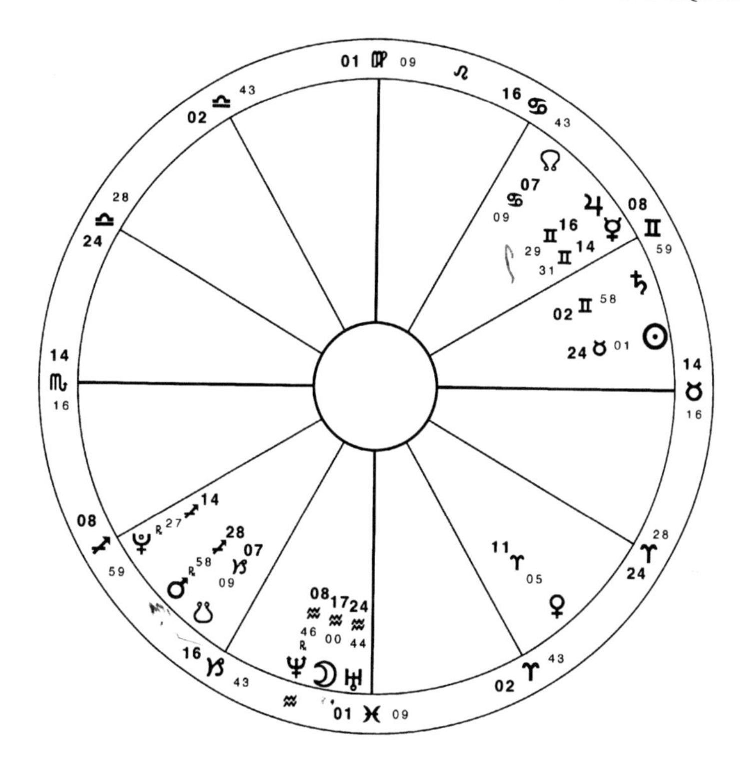

Chart 13. General health? May 14th 2001. 7.43 pm GMT. London.

attempts to rectify the situation by expelling some of this excess blood (and so there were). That the fifth house is afflicted by the presence of debilitated Venus, while the ruler of that house is that stagnant Mars, suggests that the seat of the problem is in the liver (fifth house).

While the immediate issue is the excess of blood, there is an underlying situation where there is too much heat in general. Or, rather, too much heat that is unable to pass through the system as it should. This is a common disorder, as an excess of fire is a particularly difficult matter to treat in modern society. In the workshop we take care to remain on difficult terms with certain of the neighbouring tribes, so that whenever a few of the stable-lads show signs of such an excess of heat we can provide a release for this energy by packing them off to attack the barbarians with slings and arrows. Our informants in the modern world suggest that such behaviour is no longer socially acceptable.

This is indeed a problem, as it leaves little scope for burning off such excess fire, which consequently stagnates with the most unfortunate consequences. Sport does offer something of an outlet, particularly those sports that allow the possibility of metaphorically killing one's opponent; but for all that football is famously 'more important than life and death', sport is rarely, if ever, played as if this were truly so – even at its most competitive levels. Without such intensity, the furnace that is the human being fails to reach sufficient temperature to successfully work the alchemy of transforming this surplus energy. (That the alchemist's furnace is commonly shown in humanoid form makes an important point.) Competitive sport is a useful tool, and is far better than doing nothing at all; but it has its limitations.

As does attacking the barbarians, of course: the warrior chases his ideal of Battle as assiduously, but with the same frustration, as the philanderer searches for the face that launched a thousand ships. Certain military codes – chivalry, the samurai – have striven to bring conflict as close to the ideal as possible by excluding as much as possible of the worldly dross; but, constrained by Saturn, the warrior must always fight this battle here and now, not the ideal of battle then, and this battle here and now has always its spots and blemishes. From time to time, however, the sportsman, the berserker, or even the lover may be vouchsafed the experience of being 'in the zone'. When such happiness falls is when the furnace does, like a firestorm, generate its own power; this is when the surplus energy is transformed – the solution which we seek.

THE INTERNAL KING

In a world where the bizarre concept of 'humanistic' astrology can be greeted without either laughter or, perhaps more appropriately, tears, the central purpose and greatest glory of astrology is often forgotten. We are told, on the highest of possible authorities, that astrology was given to humankind as revealed or, at the very least, inspired knowledge. That is, it is a knowledge above human station, imparted to man as a grace. If it was so imparted, it is reasonable that it was thus imparted for a purpose, and that this purpose is something more profound, and of more lasting benefit, than the ability to impress young ladies with the eternal question, 'Where is your Moon?'

The benefits of astrology are manifold; the two greatest of these relate directly to our spiritual well-being. Astrology is the finest of all glasses through which we may admire the perfect intricacy of the Divine handiwork, the better to know God. It is the finest of all tools with which we may achieve an objective understanding of our own nature and then, most significantly, do something about that nature to bring ourselves closer to God. The first of these must, we are sure, be apparent in some degree to anyone who has ever studied an astrological chart. The second is as widely ignored as is the spiritual context in which it sits.

Astrology deals with the Lesser Mysteries. That is, it is one version of the 'foundation course' which must, in some form or another, be taken before entry to spiritual life. The purpose of this course is to render us into something that might reasonably be called human, drawing us from the mechanical/animal state to which we so fondly cleave. Only when this foundation course has been completed can we embark on a spiritual path – despite the common contemporary belief that entry to spiritual life involves nothing more than stepping inside the local Mind Body Spirit bookshop. Without such integration (and we use the word cautiously, eschewing the Jungian tinge which it has now acquired) attempts to tread the spiritual path can result only in the feeding of elements of the unintegrated baser nature, which feeding necessarily has unfortunate results. This is the significance of the story of the Sorcerer's Apprentice.

To find these Mysteries in the chart, we must be willing to divest ourselves of the ragged garments of modern thought and remove robes more splendid and more befitting a man from the cupboard in which they have been locked since what is, by some perverse irony, known as the Enlightenment. Modern thinking and an astrology that has been remade to pander to a world in which the spiritual is no longer, in any meaningful sense, a current concept, will do nothing but lead us astray. As the history of the Tower of Babel makes plain, the necessities of spiritual life are not a matter of opinion; nor are they a man-made construct. Not even if the man who constructed them comes from California.

'Man alone among created beings is the link between corruptible and incorruptible things,' Dante explains, 'And thus he is rightly compared by philosophers to the horizon.' By astrologers too, for it is for this reason that our charts are founded on the horizon, and it can justifiably be said that the whole chart folds out from that point like a fan. It is on this lower 'horizon of eternity', as the philosophers to whom Dante refers have it, that the soul drops into the matter of its earthly vehicle. Hence Saturn's (planet of weight, gross matter, confinement and restriction) rulership of the first house – and also, as the god of doorways rules our goings out as well as our comings in, of the eighth, from where the soul is released again into the house of God which is the ninth.

Operative in the flesh, the soul takes on its several faculties, as shown by the planets in the nativity. It is through these faculties that we act and are known, and in each of us individually each of these faculties will be in a different condition. Some will burn brighter than others; some will be more corrupted by their position in the world. In most of us these faculties will be directed to any and every end rather than that which is their true purpose: ardour, for example, which is the faculty of Mars, is ardour yet, whether that ardour manifests in a zeal for prayer or in punching my neighbour on the nose. These various faculties, each signified in the chart by one of the seven traditional planets, are commonly seen as different colours, 'the dome of multi-coloured glass that stains the white radiance of eternity', as in, for instance, Joseph's coat or the matador's suit of lights.

The analysis of the chart will show us the condition of each of these faculties in the native. The tool for undertaking such an analysis is the close knowledge of dignity, the *sine qua non* of traditional judgement. Broadly speaking, the essential dignities and debilities show us what sort of condition each faculty is in; the accidental dignities and debilities show us how accessible it is. As an example, Mars in Cancer would show us that the faculty of ardour is in a bad state (Mars in its fall). Why is it in a bad state? Because of lack of boundaries (Saturn debilitated in Cancer). So it will be hard for the native to apply his ardour to the correct ends. It will tend to come out in unwarranted situations: a typical manifestation is in the man who has to pick fights to prove his masculinity. So much for the essential dignity. The accidental dignity will show how this unsavoury characteristic is used. If it is right on the Midheaven, we might have someone who beats people up for a living: this corrupted ardour is very apparent. If it is tucked away in the twelfth house we have a seething mess of internal resentments, but little external action. This broad outline would, of course, be further qualified by aspects and receptions with the other planets.

So far, we have a basic analysis of how the various faculties are working. This can be most salutary, allowing us to correct deep-seated misapprehensions over what is or is not constructive behaviour. Studying the receptions between the planets allows us to make interventions on the – in the modern sense of the word – psychological level. If our native's Venus (the faculty of conciliation) is in mutual reception with this debilitated Mars, he can see that spending all his time starting fights doesn't do much for his love-life, and so moderate his behaviour.

This is elementary and relates to the day-to-day business of picking a pathway through life. It is when we want more – when we wish to take the foundation course that is the Lesser Mysteries – that a dramatic next step becomes necessary. We must find ourselves a king.

The idea that we might take the hierarchy that governs the stars as the model to rule ourselves is quite contrary to the trend in astrology that takes modern fashions in politics and attempts to impose them upon the heavens. As it is much the easier to order ourselves than it is to rearrange the stars to our whim, there is, for all its lack of vogue, much to be said for this dated idea. That it is no longer fashionable underlines the immense spiritual significance of the execution of King Charles, after which astrology necessarily went into a deep decline. 'As above, so below'; so actions on the mundane sphere cannot but be reflected in our hearts (the idea that 'people get the government they deserve' being merely a loose and particularised rendition of our favourite astrological adage). That it is no longer fashionable also makes its practice that much the harder, for there remain but few of the external supports in society that would once have facilitated our strivings for internal monarchy. It must be noted, however, that the common belief that any progress can be made in this direction without exoteric spiritual practice is quite erroneous: we are not in the position to pick and choose.

To find the internal king, we need to identify 'the Lord of the Geniture'. This is the planet that has most essential dignity and is also adequately placed accidentally. That is, there is no point in selecting a planet with strong essential dignity if accidental debilities show that it cannot act: we would be better off with a less promising planet which can do something. On the other hand, we are ill advised to go for a weak planet that is accidentally prominent: just because it shouts loudly does not necessarily mean it has anything of value to say. If the planet we select is in aspect with the Ascendant, so much the better: it will be that much easier to get it onto the throne. A trap into which we may inadvertently fall, however, is in unwittingly selecting a planet that is in aspect with the Ascendant, and hence easily accessible, without consideration of its quality.

The task is to divorce ourselves from the Lord of the Ascendant and place ourselves under the power of the Lord of the Geniture. Rather than the life we are given, we choose the life we want. (Shakespeare's *Henry IV* is a study of this process, as Hal escapes from his Ascendant ruler – hence the emphasis on the sheer, first-house corporeity of Falstaff – and becomes king; that is, he becomes truly a human being and is then fit to enter on the crusade that is the spiritual life.) Thomas Taylor explains that this task, our inherent duty, is that illustrated by Ulysses' escape from the Cyclops. As that tale makes clear, the task is not easy. Even if the Lord of the Geniture happens to be the same planet that is Lord of the Ascendant, we still have the problem of shaking it awake, crowning it and convincing it that it is now in charge.

As king, it is this planet's job to steer the ship of state. Having (comparatively) strong essential dignities qualifies it to have at least some idea of where it ought to be going; having (comparatively) strong accidental dignities gives it some degree of ability to get its views across to the rest of our faculties and whip or cajole them into line. This does involve what is, of all the unfashionable ideas we have considered, the most unfashionable of the lot: discipline; but without that, nothing.

Once the king is crowned, the native cannot simply sit back and enjoy

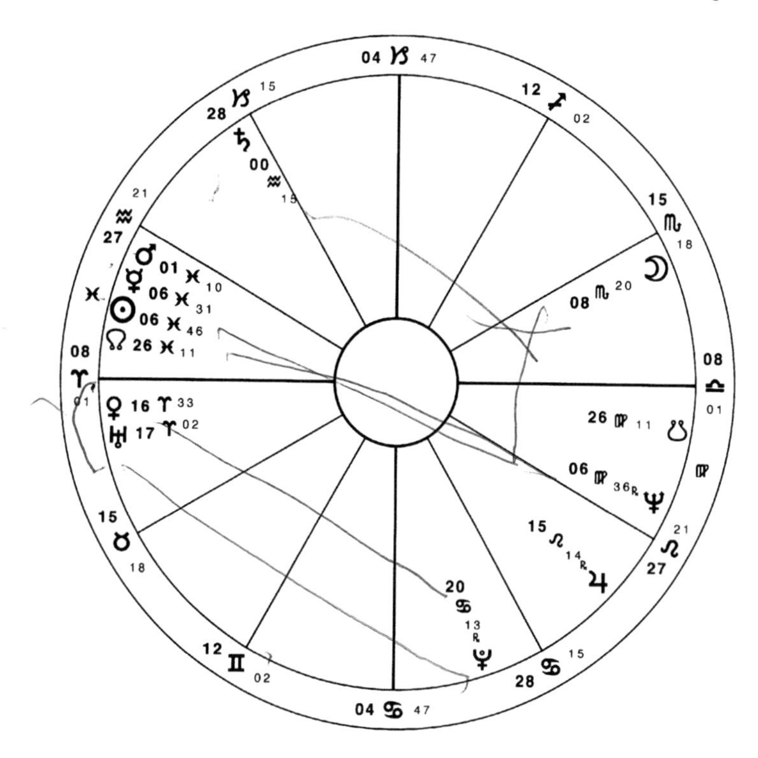

Chart 14. Johnny Cash February 26th 1932. 8.09 am CST. Kingsland, Arkansas.

a trouble-free existence – as the continuing saga of Ulysses after leaving the Cyclops indicates. The king may be on his throne, but the peasants can be as revolting as ever. So continual adjustments will be necessary. But once some order has been introduced into the soul, these adjustments are that much easier to make and that much more productive. A well-placed king finds that the sense of united direction imparted by the very fact of his reigning makes dealing with insurrections easier than the hopeless task of trying to appease every conflicting faculty under an internal democracy.

A brief example illustrates the process. As people's spiritual lives are not susceptible to public scrutiny, we will take an example from public life which does demonstrate something of the possibilities, if only in a superficial manner. The nativity of Johnny Cash shows a man with enough demons for a whole town. As the weight in his house of self-undoing, allied with detrimented Venus in the Ascendant, makes clear, there is a propensity for any amount of self-destructive fun. Entertaining it might be; wholesome definitely not: so there is an urgent need for the Lord of the Geniture to take the helm, not even to steer to the Promised Land, but just for the immediate business of keeping the ship off the rocks.

There can be no doubt about the choice of planet: Saturn is in its own sign, triplicity and terms, so immensely powerful essentially, and is accidentally strong. Saturn being Lord of the Tenth, Cash had the opportunity to 'plug into' it through his working life, or in the whole manner of the way he presents himself to others. He had the good fortune to stumble into the role of 'the Man in Black' as his stage persona. Cash seems to have done this by chance rather than design, so the peasants in his first and twelfth houses have continued revolting throughout his life. But the Lord of the Geniture, the man in black (Saturn), who 'walks the line' (Saturn); who has seen it all, done it all, and done it all often enough to know better (Saturn), weighs heavily against these proclivities and has undoubtedly saved him from the more painful clutches of his particular demons.

NEPTUNIA REPLIES...

– a word from our sensitive seer

Dear Neptunia, I am so confused; I know only you can help me. My boyfriend has always has a bad temper: he says it is because he has Mars in his first house. But I have learned to live with that. But the other night, we were going to the pictures and he didn't turn up. He said it was Saturn transiting his third cusp, so he was unable to go anywhere. What should I do?

Yours despairingly, Tracey

Dear Tracey, How I sympathise with you; I hear this kind of thing all too often. Indeed, in the wild days of my youth, when I had yet to don the garb of staid conformity, I had recourse to such odd logic more than once. I remember a particular occasion, when in my flame-red Ferrari I had been flagged down by a member of the constabulary. My explanation that I must drive fast as Mars rules my house of short journeys cut ice neither with him nor with the magistrates next morning.

What your boyfriend fails to realise, Tracey dear, is that saying 'I am furious because my Mars is such-and-such' is not different to saying 'I am furious because je suis furieux': it merely repeats the same statement in two different languages. So for him to blame Saturn for his standing you up is to blame his standing you up for standing you up. Would you turn a tolerant ear if he were to tell you, 'I didn't turn up because I didn't turn up'? I think not.

Such a logic disregards the basic tenet of astrology: as above, so below. In only four words there should be little scope for confusion. They mean what they say: what is below is as what is above. They do not mean 'What is above controls and takes responsibility for what is below'. My good friend of many years' standing, Mr. Trismegistus, has wit enough to have said that if that were what he had meant.

The planets in our charts are not external operatives, exercising their whims on our lives; they are within us as much as they are up above us in the sky. Mars does not control my anger: Mars is my anger. Saturn does not control my restrictions: Saturn is my restrictions. So when a planet transits a point in my chart, it is not an external entity dropping bombs on my life, but the unfolding of my innate – in the most literal sense – tendencies. This is when my anger explodes; that is when I fail to turn up.

So, Tracey, I suggest that the next time he comes out with such nonsense, you direct your own Mars to both ends of his anatomy, to demonstrate to him that what is above and what is below are united by a third force which controls them both.

Your caring, Neptunia

6

The Houses

THE FIRST HOUSE

In his classic text-book, the greatest of English astrologers, William Lilly, gives the vital prerequisites for any student of the celestial science. The first of these, obviously enough, is the ability to calculate an astrological chart.

Nowadays, the only ability that this involves is in knowing which button to push on a computer keyboard; so while there is still a certain value in understanding the mechanics of constructing a chart from scratch, the acquisition of this ability is no longer much of a barrier on the student's path to knowledge. All the more importance, then, falls on the second of Lilly's necessities: the sound knowledge of the meanings of each of the twelve houses of the chart.

This knowledge is essential. It is this, and only this, that tells us where to look within the chart to find the information that we require – and if we look in the wrong place we shall inevitably unearth the wrong information. It is probably true to say that the greatest number of errors in astrological judgment are made simply through looking at the wrong house.

It is worth taking some trouble, then, to consider and understand the meanings of the various houses. This is true no matter which branch of astrology we wish to practice, whether it be natal, horary, electional, mundane, or any of the minor branches.

Before we examine the houses, it is as well to cast an eye over the various methods of deciding where the boundaries of these houses fall. It is often claimed that the Equal House system of house division that is so popular today is the most ancient, but this betrays a misunderstanding of the ancient method. The old system involved taking each sign as one house – but this was the whole of that sign, not just a part of it as with the modern system. That is, if the Ascendant were in Taurus, the whole sign of Gemini would form the second house and the whole of Cancer the third, and so forth. It made no difference to these other houses at which degree of Taurus the Ascendant fell. This is unlike the Equal House, which repeats the degree of the Ascendant in the cusp of each of the other houses.

This system of whole-sign houses is still the main system employed by

Vedic astrologers, and in skilled hands it gives results of great accuracy. More common in the West, however, are the various unequal systems of house division.[1] The most familiar of these is the Placidean.

Just as it is commonly claimed that the Equal House is a system of great antiquity when it is not, so it is often said that the Placidus system is comparatively modern, when it has been around far longer than is usually realised. The error arises through its name, as Placidus de Tito, on whom it has been fathered, died as recently as 1688. Although he popularised the method, it can be traced back to the first millennium.

Perhaps the most remarkable thing about Placidus is that he ever managed to popularise anything, as his book, *Primum Mobile,* is dull to the point of unreadability. This may not appear to set it apart from certain other classic astrological texts, but Placidus' mastery of torpor gives his work the crown.

The major rival to this system was that of Regiomontanus. This takes its name from the pseudonym of Johann Muller, one of the greatest of mathematicians. His method works excellently for horary charts, while Placidus is verifiably better for most other purposes – not, it must be noted, through the subjective decision 'That sounds more like me', but through prediction of events by directions to house cusps, which are, of course, the variable between the different systems.

Almost all of the hundred-plus house systems that have been used by somebody-or-other somewhere are rooted in the Ascendant, which is the eastern horizon, the dividing line between earth and sky. This is the cusp of the first house, and as its role as foundation stone of the chart implies, it is of crucial significance. The Ascendant contains within it as potential all that exists in the rest of the chart.

In the ancient philosophical texts, such as the pseudo-Aristotelian *Book of Causes,* the horizon is accorded great significance. It is seen as the place where the soul assumes its human incarnation. It is this which gives the first house its major meaning as the person and body of the 'owner of the chart': the querent if it is a horary chart, the native if it is a birth-chart, the city or country if it is a mundane chart.

[1] That these systems appear unequal is an illusion. Each Placidus house, for instance, is equal in that it signifies two hours of real (i.e. in contrast to clock) time; each Regiomontanus house is equal in that it subtends thirty degrees, but of Right Ascension rather than the more familiar Celestial Longitude. It depends what we mean by 'equal': is my house equal to yours because it has the same number of bedrooms, or because I can sell it for the same price?

In modern astrology it is customary to see the whole of a birth-chart as being 'this person'. In the traditional method, however, this is true only to a limited extent. Yes, the whole chart shows that person and his nature; but we can also pick out the native as one individual moving around within the chart, and locate there the other people who populate his life – brothers, sisters, partners, parents, enemies, and so forth. It is the first house and its ruler that show us the native.

Most of our judgment of the native's appearance will be taken from the first house and its ruler. This is quite contrary to – and rather more accurate than – the common attribution of appearance to the Sun-sign. Of particular importance here are planets in the first house and any planet that is in close aspect to the ascending degree. Such a planet usually has the predominant voice in determining physical appearance.

Any planet in the first house will also be our first choice as significator of the native's manner, or of his general mode of behaviour. There will be a world of difference between the native with a dignified Venus in the first and that with a debilitated Saturn – most of us would much rather entertain the former to tea!

Saturn in the first house is by no means always bad news, however. Saturn gives strength (in the sense of endurance) and resilience, while the first shows the body. So provided Saturn has some sort of dignity his placement there is a good testimony of a robust constitution and long life.

While the first house shows the body as a whole, the body is also shown by the whole chart, working downwards from the first at the top to the twelfth at the toes. Specifically, then, the first shows the head. It is a fair bet that anyone you know with a scar or patches of pigmentation on his face has Mars in the first house.

This association with the head is something that the first house shares with Aries. But, in an important difference from modern astrology, the traditional association of signs and houses in this manner begins and ends with parts of the body. The 'Alphabetical Zodiac' that equates Aries with Mars with the first house, Taurus with Venus with the second and so on has no place in the traditional view of the craft. The association of planet with house in this pattern is never used in traditional astrology, as it makes no sense in the light of the inner meanings of the various houses. The house/sign comparison is indeed most valuable: but only when used for medical or descriptive purposes.

As the Ascendant is the first of the houses it is Saturn, the first of the planets, that belongs there, not, as the moderns would have it, Mars. Saturn is the god of doors, ruling our goings out (he also rules the eighth house) and our comings in. He rules the body – hence his attributes of weight and solidity. He rules all boundaries, such as the skin, within which our soul is now encased. And, of course, the greatest of all boundaries is that which separates what is alive from what is not alive, and it is this dividing line that is shown by the Ascendant.

Although Saturn is the planet associated with the first house, and as we have seen, he can be fortunate when placed there, it is Mercury that actually gains in dignity by being there. The first is the house of Mercury's *joy.* The connection here, on a superficial level, is between the head and the reason and loquacity that is placed within the head. On a deeper level it takes us back to the idea of the first being the boundary between the the immaterial and the material, the doorway into incarnation. So Mercury's joy here gives us a picture of the Word (Mercury) made flesh (first house).

Also redolent with meaning is the association of the first house with our name. Name is something of greater significance than is commonly accorded it today. Our name images our essence, the deepest part of our nature, that which makes me me and you you. That our given name is thus associated with the innermost part of our nature is why certain tribes have a taboo against revealing personal names to strangers, feeling that this knowledge will give the stranger power over them. When progressing the birth-chart, it is typical to find the sign on the Ascendant changing when the native changes her name. This is, indeed, one of the most reliable indications for rectifying a chart – if we are fortunate enough to have such an event with which to work. In the horary questions which occasionally arise on the merits of changing a name we will look at the ruler of the Ascendant and see if it is happier where it is, or where it is going. This will show whether or not the change is beneficial.

While Mercury is stronger when in the first house – the reason, we might say, being in the head, where it belongs – the Moon does not like being there at all. The Moon and Mercury form one of the many pairs into which the seven planets can be divided, and if one of them is happy somewhere the other will usually not be.

The Moon is our emotional nature, and as such it does not fit well into environments where our reason is content. It is also the significator of the psychic substance of the soul; as such, its unhappiness in the first house mirrors the soul's ambivalence at finding itself in the body. On the one hand it is excited at the range of experience that now lies before it; on the other it is horrified at the weight and constriction that this incarnation involves.

In a natal chart, the Moon in the first house feels herself far too much exposed, as if the protective veneer that keeps our finer feelings from the vicissitudes of life is too thin. This placement is a strong sign of a changeable nature. This is not changeability in way that, for instance, Gemini will change, driven by the endless curiosity that it holds. This is changeability from over-sensitivity. The Moon in the first might remind us of those occasions when we have been sitting far too long in one place: we keep changing, shifting from one buttock to the other as we try to make ourselves comfortable. This is very much the condition of the Moon in this most uncongenial of positions: try as it might, change as it may, this sensitive nature cannot find a situation where it feels at home.

In an electional chart the first house has prime importance. It shows the enterprise itself, and it also shows the instigator of this enterprise, who is usually the person for whom the chart is being constructed. It is vital, then, that the first house is made as strong as possible: we want to put our man in the best position.

It is important also that the nature of the first house should reflect the nature of the enterprise. If it does not, it is as if we were making a start at an inappropriate time. If we want the job to be over and done with as quickly as possible, it is unsuitable to begin when a fixed sign is rising: that will give delay on all occasions. If, however, we are building for the long term, a fixed sign is exactly what we need.

The particular field in which our enterprise lies should also be shown by the sign on the first cusp. If we are opening a library, for instance, we would want an air sign on the Ascendant. If it were a fire-station, we would go for a water sign; if a garden-centre, we would prefer earth.

In a mundane chart, the first shows the general state of affairs in that place, whether it be the chart for a town or a country. It also has particular reference to the common people. If, then, we were examining the Aries ingress chart in a particular place to find out what the weather would be like, we should look at the first house.

Finally, we must always pay especial attention to whatever fixed star is rising over the Ascendant. In no matter what sort of chart we are considering, the influence of such a star acts much like the wording on the title-page of an old play. If we see the word 'tragedy', we know we are in for trouble; if we see 'comedy' we know everyone will live happily ever after. Indeed, it is not inaccurate to view the first house as being the 'title-page' of the chart. Everything else in all the other houses just expands and amplifies what we find here.

THE SECOND HOUSE

The great point of confusion that is the first stumbling-block faced by any beginner in astrology is the matter of which way round the houses go. It can seem quite illogical, as some things in the chart seem to be going in one direction, while others go in the opposite. Why the houses should be numbered in an anti-clockwise, rather than a clockwise, sequence can remain a mystery for a long time!

The apparent contradiction is caused by our dealing with two separate kinds of motion. There is the rapid motion relative to ourselves, easily visible by watching the Sun move round the sky. This motion starts in the East and moves towards the West, completing a full circuit in approximately one day. This is the motion created by the Earth's daily rotation and is called 'primary motion'.

The other, or 'secondary' motion is the movement of the various planets through the signs of the zodiac. This is far slower than primary motion, and proceeds in the opposite direction. This motion is the consequence of the planets' – including the Earth's – movement around the Sun.

We may take the Sun as an example. By primary motion it travels all the way around the Earth every day. But by secondary motion it travels one degree in the opposite direction to this each day – hence its gradual movement from one sign to the next. We might see the planets as a number of flies walking round a plate. The plate is revolving fast in one direction (primary motion) and carries the flies along with it. But they are walking in the opposite direction to the plate's own movement (secondary motion). Except, of course, when every now and again one of the flies turns retrograde.

When looking at the chart, the primary motion is in a clockwise direction. So planets rise in the East, at the Ascendant, and then traverse the houses in reverse numerical order: first the twelfth, then the eleventh, tenth, and so on. The signs of the zodiac, meanwhile, and consequently the houses (remember that the signs are the 'celestial' houses while the houses are the 'mundane' houses) through which the planets take their secondary motion follow in an anti-clockwise direction.

There can be few student astrologers who at some time or other have not drawn a chart upside-down, becoming confused by this conflict in direction and concluding that the second house should sit above, rather than below, the first. But the whole picture of the mundane houses succeeding each other in regular order like the spokes on a wheel is in itself misleading.

Rather than being a wheel, the chart is in fact an assemblage of four groups of three houses. Nor even are these groups divided in our usual way of seeing the four quadrants of the chart: the quarter between Ascendant and Midheaven being one quadrant, that from the MC to the Descendant being the next, and so forth. The four groups are centred around the four angles of the chart (Ascendant, MC, Descendant and IC), each of which is flanked by two other houses. This pattern is a good deal clearer in the traditional chart-form, which is square, not circular, and is divided into four triangles, each made up of three smaller triangles, grouped around a central box.

The four angles are the structural key to the chart, like the main beams in a roof. The other houses, as it were, lean on the angles for support – like the rafters coming off the main beam. This is why the angles are 'there' in a much more definite fashion than the other houses: they are more real, more solid, more tangible, and a planet in one of the angular houses has more power to act.

The houses following each of the angular houses in an anti-clockwise direction are the 'succedent' houses, so called because they follow or succeed to the angles. These are houses two, five, eight and eleven. But the succedent houses are the last of each group of three, not the centre of it as is common to think. The cadent houses (three, six, nine and twelve) do not follow on from the succedent, but fall away from the angles.

This is why they are called 'cadent' (literally: falling). Indeed, rather than all the attempted reforms of astrology incorporating everything from asteroids and hypothetical planets to micro-aspects and random symbology, the most useful reform that astrologers could undertake would be to practice their art in English! We suffer inordinately from our persistence in working with terms that are lifted straight out of various foreign tongues without being translated. If we spoke of 'falling' rather than 'cadent' houses, their significance would be much clearer – as also would be the simple fact that planets in them have trouble acting: you can't do much if you are falling. Similarly, referring to planets 'glancing' at each other, rather than making aspects ('aspect' is glance in Latin), would clarify all manner of confusions about what is or is not truly an aspect.

Perhaps more significantly still, we might consider the names of the fixed stars. That a star is 'the tail of the goat' or 'the eye of the dragon' tells us a great deal about its nature and effects. This information is lost to us if the name is hidden in Latin or Arabic.

The meaning of the second house

Our chart, then, can be seen less as a flat circle than as a group of four mountains, each of which has a smaller hill on either side of it. In our journey of exploration round the houses, the first of these hills at which we arrive is the second house, the house which follows and supports the mountain that is the first.

It is this sense of supporting that gives the second its prime significance in the chart, as it shows the resources of the first house. It shows what the native or the querent has at his disposal to sustain him in his life, or in the situation of the horary.

The term 'second' is found in exactly this sense if we think of a duel. My second is the person who helps and supports me, loading my pistol and bandaging me up afterwards. One meaning of the second house is anyone who acts in this capacity. So when I go to court, my lawyer is shown by my second house. At work, I would assign my employees to the sixth; but if I wished to distinguish my right-hand man, or my PA, perhaps, I would give them the second.

This house's most common meaning is money. This is money in any form, not necessarily currency. It is a common error to ascribe shares to the eighth house; but my shares are still my money – just money in a different form. So they are still second house. Anybody else's money is shown by his or her second house.

In any question about profit, it is usually this other person's money in which I am interested. There isn't much point my asking about my own money: I am all too aware of the condition of that! So to answer a horary question about whether I will make money, I need to connect the ruler of the other person's money with either my own significator (the money comes to me) or the ruler of my second (the money comes to my pocket).

Once I have established a connection with the other person's money, I can judge how much money I will get by the strength of its significator. I hope to find this in as strong a condition as possible: the stronger it is, the more money comes my way. The nature of the aspect will show how easily I get it: by conjunction, trine or sextile it will come without problem. With a square I will have to chase it and should expect delays. By opposition, it will probably not repay the effort I expend obtaining it.

What I do not want to see in a question like this is an aspect between the ruler of the second and my own significator. That shows my own money coming to me, and that is no use at all. Even if I am asking about money that I have lent, for the purposes of the question it is the 'other person's money' that I want.

In most cases, this will be shown by the eighth: the second from the seventh. If I am waiting for my wages, I would look to the eleventh: the second from the tenth. So also if I am expecting money from someone powerful – 'a present from the king', in archetypal terms. William Lilly echoes the Bible when considering questions on recovering loans from powerful people. No matter what the chart says, he advises, if the person who owes you money is more powerful than you are, forget it!

The eleventh is also the house I would look to if hoping for a lottery win. But any other question of gambling is a second/eighth house matter. It is *not* a fifth house question: this may cover having fun, but if my aim in betting is to show a profit, it reverts to the second and the eighth.

Apart from money, the second shows my other possessions. Astrology retains a definition of possession that is no longer, unfortunately, current in the western world. I can possess something only if it is inanimate and if I can move it around. So my house or my land do not fall within the second house. My car, however, does. The whole point of a car is that I can move it around, so the common modern ascription of cars to the third is quite wrong. The third (or the ninth) may show the journey, but it does not show the car itself.

At a deeper level, the second shows our self-esteem. Dante gives the reasoning behind this connection. As he explains, if we think well of ourselves we will think well of our possessions. This is not the same as the materialistic idea that if I have a flash car I will value myself more; it is, in fact, the exact opposite. It suggests that if I value myself highly, I will think highly of whatever sort of car I happen to possess. We might remember the benevolent influence of Jupiter on the second, which makes this reasoning clearer.

For, being the second of the planets, Jupiter is natural ruler of the second house, as Saturn is of the first. Not, it must be stressed, Venus. The equation second house = second sign = ruler of second sign is wrong. Jupiter is the planet of wealth, and so fits the second well. When studying a natal chart to find out how wealthy the native will be, Jupiter must always be considered together with the second house and its ruler. So also in any general horary question on the same subject, such as 'Will I ever be rich?"

The third point to take into account when looking for wealth in the nativity is the Part of Fortune. The inner significance of this is on the level of our spiritual life, but the external meaning helps indicate how much money we are likely to have, and where it is likely to come from. When judging the condition and placement of Fortuna, we must always look also at its dispositor.

The French astrologer, Jean-Baptiste Morin, wrote a brief but profound study of the houses. He links the second to the sixth and the tenth houses in what he calls 'the triplicity of gain'. The tenth, he says, is the most noble of these as it relates to immaterial objects of gain, such as honour and the position indicative of such honour. Our standing in the world.

The sixth he says is 'material and animated', showing the living things at our disposal: our servants and small animals. The second is the inanimate things at our disposal: our money and other treasure that we have gained by our own actions.

Morin's attempts to purify astrology are contained in his massive *Gallic Astrology*, of which only a small part has been englished. One spurious modern book claiming to explain his methods has even managed to relocate him a few hundred miles further north, fathering upon him the previously unknown 'Gaelic Astrology'! So much for accuracy.

As the second shows our resources, it also shows our food, the most important of all material resources. As a bodily part, it shows the throat. These two meanings go together, showing the perfect congruence of the astrological model of the cosmos. As the second is the throat it is also, by extension, what we eat.

In a recent horary the querent asked about an impending throat operation. The chart showed Mars, natural ruler of surgery, applying immediately to conjunct the second cusp: as perfect a picture of a throat operation as might be wished. Similarly, the chart will make clear any aberrations regarding what we do or do not eat, and the second is the place to look in order to find them. Afflictions to the second cusp, especially from restrictive Saturn or the South Node, which operates in much the same way as Saturn, are typical of anorexia. Jupiter or the North Node, although they are both usually regarded as benefics, can show over-indulgence, while Mercury is likely to show fads and fussiness. As with all indications in the natal chart, however, it must be stressed that these cannot be read alone. There is nothing more certain to lead to error of the most gargantuan proportions than dipping into the chart, fishing out isolated testimonies, and judging them without reference to the rest of the picture.

This is equally true of the indications for wealth: whatever is shown of either good or bad will rest as potential if not activated by other factors within the chart. I may have the ability to earn millions, but if I am so lazy that I never apply this ability, it will never manifest in my life. This is the great caveat behind the study not only of the second, but of all the houses of the chart.

THE THIRD HOUSE

Of all the houses of the astrological chart, it is probably the third that arouses least interest. In most birth-chart readings it will be quietly skated over, as the astrologer can usually find nothing there that warrants closer examination. To treat it thus is to do it a disservice, however, as it plays a pivotal role in the true assessment of any nativity.

The immediate meaning of the third is as the house of brothers, for it shows the native's – or the querent's, if it is a horary chart – siblings. As an extension from this, it also shows any relatives on the same generational level as the native. That is, your cousins, being the children of your parent's siblings, belong here, even if they are thirty years older than you; but your uncle does not, even if he is your own age.

Studying the relative strengths, placements and receptions between the rulers of the first and third houses in a birth-chart will show whether the native gets on with his siblings and which of them has more success in life. It is not unusual, unfortunately, to find adverse indications revolving around either the second house (the native's money) or the fifth, which is (second from the fourth) the house of the parent's money.

We have various ways of distinguishing which of our siblings is which. Mars is the natural ruler of brothers and Venus of sisters, so we can look at these two planets for further clarification. Or we can take the third house to show the eldest sibling, and third from the third to show the next, and so on. By taking the third from the third we are regarding the second sibling as the brother or sister of the first.

Alternatively, we can use the triplicity rulers of the sign on the third cusp. It is common to regard each of the four elements as having two triplicity rulers, one by day and the other by night. Some of the early astrological texts, however, use three. This is not to suggest that either schemata is right or wrong: they are different tools with different uses. The triplicity rulers are:

Fire: Sun, Jupiter, Saturn
Earth: Venus, Moon, Saturn
Air: Saturn, Mercury, Jupiter
Water: Venus, Mars, Moon.

When judging the third house we would take the first of these rulers to show elder siblings, the second ruler to show the middle siblings and the third to show younger siblings. The list above gives the triplicity rulers for a day-time chart – a chart with the Sun in the seventh to twelfth houses; the first two rulers in each case reverse for night charts. So fire by night, for example, is ruled by Jupiter, Sun and Saturn.

Suppose, then, the third of these rulers were in its exaltation and in strong mutual reception with the Lord of the Ascendant, which was in its own triplicity. We would judge that our kid brother was going to be rather more successful than ourselves, but would be happy to do whatever he could to help us when the need arose.

Extending the idea of people on our own level, the third shows our neighbours. This is meant in a strict sense as the people who inhabit houses close to our own, rather than the sense given in the gospels as anyone whom we happen to meet. This is because our neighbour's house in the chart (the third) is, in the most literal sense, next-door to our own house (the fourth). In any horary chart about house-purchase we must pay close attention to the third house: no matter how excellent the house itself might be, we can still be miserable there if we have crazy neighbours.

Nowadays the third is known primarily as the house of communication, and this is indeed one of its more important roles. It covers our speech – although the instrument of speaking, the tongue, is found in the first house, which rules the head. It shows letters, phone-calls, rumours and messages. There is something of a grey area over where we locate messengers. Ambassadors and messengers belong in the fifth, the distinction being that they have some power to negotiate on our behalf, while a third-house messenger just delivers the message and then shuts up.

It is to the third house that we look if we wish to determine if and when a certain package or letter is due to arrive. In this case, it will usually not be our own letter with which we are concerned, so we would turn the chart. Most often, we would look at the third from the seventh, as a letter from 'any old person'. If it were from our child, we would take the third house from the fifth, or if from our mother, the third from the tenth. We would hope to find an applying aspect between the ruler of this house, signifying the letter or package, and either the ruler of the Ascendant, the Moon, or the Ascendant itself. The physical arrival of a person or object is one of the few instances where an aspect to a house-cusp is enough to give us a positive answer; most charts for questions on this issue do in fact show aspects from the significator to the first cusp.

Another frequent third-house horary question is whether a certain piece of information is true or false. The method of judging this is different from most other kinds of horary, as we are not looking for an aspect to show something happening, but seeking to find a certain sense of solidity in the chart. It is almost as if we were banging the chart on the table to see if it is real. To show the information as true – whether this is for good or for bad – the chart would have fixed signs on the angles, especially on the Ascendant and Descendant. The ruler of the Ascendant should be in a fixed sign and an angular house. The ruler of the third should be in a fixed sign and an angular house – and a fixed sign on the third cusp is always helpful. Both the Moon and its dispositor should be fixed and angular.

It is a rare chart in which all these elements would be in place; but given a fair number of them, we would judge that the information is true. If we had an absence of fixity and a preponderance of planets in cadent houses, we would know it must be false. The fixed signs and angular houses share the sense of solidity and reliability for which we are looking. Once we have decided that the information is true, we can then look at the condition and relationships of the planets involved to find out if it is in our favour or not.

That the issue of truth is found in the third house demonstrates the importance of its placement in the chart, for the third is opposite the ninth, and serves as a reflection of that house – and the ninth is the house of God, or of Truth. This reflection is exemplified by the Moon having its *joy* in the third house and the Sun in the ninth. The Sun is the visible symbol of the manifestation of God in the cosmos, and the Moon, of course, shines only by light reflected from this source. That is, the Sun (and, by extension, the ninth house) is the source of all truth in our world, while the Moon is what reflects this truth to us. Much significance is given in the ancient texts to the mottled nature of the Moon's surface. To our eye, the Sun appears as a homogenously brilliant disc; the Moon, however, has patches of light and patches of darkness: not all that that we receive from it can be relied upon.

It is this axis of truth and – in every sense of the word – knowledge that runs from the ninth to the third house that renders the third so significant. While the other points which we may draw from the chart will tell us of the nature of the person, and of the various idiosyncrasies with which he is blessed, we need to be able to determine how this nature and these character traits will be used in his interactions with the world. In simple terms, is he a goodie or a baddie?

If he is to be a goodie, his nature must, to greater or lesser extent, conform to the pattern of absolute goodness that is the Divine, shining into the chart from the ninth house. So we need to see the ninth house, its ruler and other associated planets in reasonable condition. This would indicate that there is an accessible well-spring of goodness within this person. But merely possessing this well-spring of goodness does not necessarily mean that the person will act in a good way: there's many a slip 'twixt cup and lip. While the inner nature may aspire to the highest standards of behaviour, there are any number of reasons at any number of different psychological levels why the actual standard of behaviour can fall far short of these aspirations. Whether or not this happens is shown by the nature of the third house and its connection with the ninth.

Once we have established that the person means well by finding the ninth house in good order, we need to find a means of conducting these good intentions into his daily life. This will be shown first by finding favourable indications in the third house and its ruler, and then by finding – whether by aspect or reception – some connection between the third and the ninth. Failing that, some sort of influence over the third by any planet in strong essential dignity is a good deal better than nothing.

For while the third is our house of 'short journeys' it has a significance far wider than our daily bus trip. The top end of the third/ninth axis shows our long journeys, the type of which is our life-time's journey to God. The bottom, mundane, end of the same axis is all the routine journeys that make up our daily round, the type of which might be our journey to the kitchen to make a cup of tea. It is in these 'short journeys' – by which is meant all the routine business of our daily lives – that whatever ideals we may have on a ninth-house level find their manifestation.

This idea repeats what we are told in the Bible that *a man is known by his words:* it is the third house which gives the outward manifestation of what is happening in the ninth. The quality of the routine journeys that make up our daily life reflects exactly the place we have have reached in the life-long pilgrimage that is our ninth house. *A good man draws what is good from the store of goodness in his heart; a bad man draws what is bad from the store of badness. For a man's words flow out of what fills his heart.*[2]

While this is the significance of this axis on a deep level, the same axis manifests in other ways too. Our long journeys may all signify our journey to God, but they still include a week in Benidorm. It is not so much the length, but the specialness, of the journey that sets it apart. I may commute to New York three times a week: that will still be a routine, third-house journey. I may spend a weekend in a local resort, but that is a special journey, and so belongs in the ninth. The derivation of the word holi-day makes this plain.

Our minds travel as well as our bodies, so the third and the ninth show different levels of knowledge. The 'three Rs', the basic knowledge that we need to navigate our way through daily life, belong to the third house; all forms of higher learning to the ninth. So my primary school would be shown by the third, my secondary school by the ninth.

Similarly, our basic ability to communicate is shown by the third. A debilitated Saturn on the third cusp, for instance, might show a speech impediment; the ruler of the third very swift in motion and in a double-bodied sign could show someone who never stops talking. If the third is our ability to communicate, the question of whether we have anything that is worth communicating must be referred up to the ninth.

In the body, the third house shows the shoulders, arms, hands and fingers. Both this and the other meanings of communication and movement indicate the alignment of this house with Gemini. But we must be careful not to extrapolate from there by holding that Mercury rules the third: in the traditional model of the cosmos, it does not. For all that many of the meanings of this house do fall into his natural provenance, it is not Mercury, but Mars, who holds sway here.

Why Mars? The immediate answer to this is that Mars follows in the customary order of the planets after Saturn and Jupiter, who rule the first and second houses. But the assignment is by no means random. For it is Mars that gives the desire to communicate, the desire to make all the short journeys of our life. Mars shows the nature of this house at a deeper level than Mercury, giving the underlying 'why' behind all its meanings.

Even though Mars is a malefic, it usually behaves itself reasonably well when placed in the third house. It is as if the cosmos has found it a

[2] Luke 6 : 45

worthwhile job to do, so it is too busy doing that to cause any trouble. The Moon, meanwhile, is the planet that is most happy in the third, as it is the house of its joy. Here, we see all the busyness and speed of motion that is the Moon's. If she is placed here in a natal chart, Lilly tells us, especially if in a cardinal sign, 'it's an argument of much travel, trotting and trudging, or of being seldom quiet'. Seldom quiet, that is, either with the feet or with the tongue.

THE FOURTH HOUSE

As we continue our tour of the houses, our itinerary brings us to that seat of much contention, the fourth. No self-respecting astrologer is without a copious number of reasons for beating his fellows about the head, and this house provides more excuse than most, for it is where we find our parents.

The question over which astrologers are so prone to fall out is 'Which parent?' Each possible variation has its partisans: some plump for Mum; some prefer Dad; some sit on the fence by ascribing it to 'the dominant parent'. This last point of view helps us not at all – for how can we determine which is the dominant parent? The absent father, for example, can affect us more by his absence than he ever could by his presence; while if, as an extreme case, Dad died before we were born, yet we still carry his appearance and temperament, who is to say that he is not the dominant party?

The idea that the fourth might show Mum is rooted in the modern concept of the Alphabetical Zodiac, which equates signs, houses and sign rulers in a way unknown to astrology's long tradition. Because Cancer is the fourth sign, the argument runs, it must be the equivalent of the fourth house, and so its ruling planet, the Moon, must be the natural planet of the fourth. As the Moon signifies the mother, so the fourth house must show our own mother.

This is a persuasive argument, so far as it goes; but it is contrary to sound astrological thinking. As we have seen, the tradition that has served astrologers so well for thousands of years takes the outermost of the planets, Saturn, as natural ruler of the first house. Jupiter is given to the second, Mars to the third – and the Sun to the fourth. The Sun, of course, has nothing to do with Mother: it is the natural ruler of men and of fathers. So the traditional model gives the fourth house to the father.

Beyond just father, however, this is the house of both our parents, as, at the bottom of the chart, it is the root from which we spring. If we have a general enquiry as to the state of our parents, we would look to the fourth. As soon as we wish to differentiate between them we would take the fourth for Dad and the seventh from the fourth – the tenth – as Dad's partner: Mum.

In the same way, the fourth is our ancestry in general, our personal heritage. In a broad manner, all our ancestors, recent or long-gone, are located here. But again, if we wish to distinguish between them we would find other houses to signify them as individuals. My father's father, for instance, is shown by the fourth house from the fourth: the seventh.

On a still wider level we find here not only our personal but also our national and cultural heritage: our homeland, and all that we acquire from those that have gone before. If we wish to locate our home country in a chart, perhaps as part of an enquiry as to whether we shall prosper best in our native land or elsewhere, it is here that we shall find it. Foreign climes will then be shown by the ninth. This can be problematical in horaries when people have lived for a long time in another country: where is 'home'? The best course is to ask which of the two countries the native himself regards as home.

As the base of the chart, the fourth house represents what is on or under the ground. It is here that we find cities, farms, gardens and orchards. If a horary question concerns the purchase of land, we look to the fourth to see the condition of this land. If, for instance, we wish to farm it, the Sun in Leo in this house would – no matter that it is so strong – be a powerful negative testimony: the burning Sun in a barren sign would show that the land is parched and useless. A favourable Jupiter or Moon, on the other hand, would bode well for the land's productivity.

It is this concept of the fourth as the base or the solid root of the chart that gives it its rulership over those of our possessions that do not fit into the second house: those that we cannot move around. So, apart from land, whatever property we own or rent is found here. In a horary for buying a house, the ruler of fourth will tell us as much as a surveyor's report about the building's condition. If it is strong, all well and good. If it is weak, we find out why it is weak: weak in a water sign, for instance, we would check the damp-course; weak in a fire sign, we would suspect that the heating doesn't work, or that the plastering is not sound.

What we seek to find in a horary on house purchase is a balance between the condition of the fourth house ruler, showing the property, and the condition of the ruler of the tenth, which signifies the price. We hope to find them equally strong or, better still, weighted in our favour. This would show that the price is fair, or that we are getting a bargain. It is perfectly possible that we might be prepared to pay more for the house than it is worth: we might like it so much that we feel it is money well spent; but we should at least be aware that we are paying over the odds.

In such a question the ruler of the seventh shows the seller. If this planet is weak, it may be an indication that he cannot be trusted – especially if it is in the twelfth house, where we are most unlikely to find any honesty. A strong connection between his significator and another planet should alert us to the possibility of him cutting a more favourable deal behind our back. An aspect between his planet and our own is testimony that the deal will go through, though we should not expect to find an easy aspect. Oppositions, which usually carry strong negative connotations, seem to be the norm in these questions, reflecting the extreme and unnecessary difficulties which are typical of house purchase. Squares, which are themselves not easy, seem to be as good as it gets.

If the property is being bought to let out or to renovate and then sell on, we must also look to the fifth house. This is the second from the fourth: the property's money, or the profit which we can make from it. No matter how poor the condition of the fourth house, a strong fifth ruler may be an argument for going ahead.

The idea of the fourth house, the base of the chart, being associated with the ground also explains its connection with lost objects. It covers, as Abu Ma'shar tells us, 'Every matter that is hedged around and covered over.' This is meant in a fairly literal sense, as lies are more properly the business of the twelfth; but it does contain the reason why it is here that we seek for mislaid goods: the essential image of a mislaid object is something that has been put down and covered over with something else. In practice, however, we find that the distinction between lost and mislaid objects is overstated. In searching for missing objects through the chart we can usually look to the second and the fourth, taking whichever of their rulers best describes the object in question as its significator.

Another thing that is 'hedged about and covered over' is a mine: our ancestors were as keen on seeking their fortune in mines as we are in seeking ours by winning the lottery – and the fourth is where we find the native's potential for making this happy dream a reality. Mines are here because they are under the ground; the lottery is an eleventh-house matter: pennies from heaven.

In some horary questions the final outcome is of singular importance, notably in court cases, where the final outcome is the verdict, and in medical issues, where the final outcome is the bottom line of whether or not the patient will recover. This is shown by the fourth house and its ruler.

In a court case both parties get the verdict. One is usually happy with it while the other is not. In the horary chart the ruler of the fourth is more like a prize: whichever of the parties' significator links up to it first wins. Its strength and the receptions between it and the significator will show us how happy the winner is with the result. Even if our man wins, the lord of the fourth might still afflict his second house: he wins, but is not awarded costs, perhaps, so he still comes out with his finances weakened.

While the nature of the ruler of the fourth and the aspects made to it will show us what the outcome of an illness is likely to be, in medical matters the fourth house itself takes on a most malefic tinge. In medical contexts 'the end of the matter' is taken in the most literal way: the grave. This is, of course, one testimony only and must be read with the rest of the chart: if the question is 'When will I get over this cold?' we would not need to investigate the possibility of death.

Morinus links the fourth with the eighth and twelfth houses, with which it forms a grand trine, as a malefic triplicity involving imprisonment and one's secret enemies (twelfth house) and death (eighth). To fit his scheme he takes the fourth's connection with ancestry in the most negative of possible ways, seeing it as our inheritance of original sin, and finding nothing in family roots except the sorrow of seeing our parents die. In this he seems to have fallen foul of that most pernicious of astrological traps, the temptation to twist astrology to fit preconceived ideas.

In contrast, if we look back to the earlier writers on astrology, the fourth is seen as a most positive house. Apart from what should be the obvious fact that this is the house of our parents and without our parents we have no existence – and reasons do not get much more positive than that! – reference is made to the age-old concept of the Wheel of Fortune. This is the image of a huge wheel, to which we are all attached, turned by the hand of a blind woman, known as Chance, or Fortune. When related to the astrological chart, the top of this wheel is the tenth house and its nadir is the fourth.

In this, it mirrors the primary motion of the planets, their daily revolution around the Earth that is the visual product of the Earth's rotation. When a planet is in the tenth, it is at the top of the Wheel of Fortune. The tenth is the house of glory and success. The early writers, however, notably the Roman poet Manilius, see the tenth as unfortunate, for once you have got to the top there is nowhere else to go except down. The fourth, on the other hand, is happy for just the opposite reason: it is the lowest point in the chart – but once we have plumbed the depths we can be sure that things cannot get any worse, and there remains nowhere to go except up.

THE FIFTH HOUSE

The fifth house is the house of pleasure; yet this does not stop astrologers breaking each other's heads as they debate with their traditional vehemence the exact boundaries of what belongs here and what does not. Unfortunately, the terms of this heated debate usually owe much to the personal morality of the astrologer and little to any sound astrological thinking.

The seventh house of the chart is the house of the marriage partner. But what about the person with whom we are involved, but who is not our spouse? The late-Victorian and Edwardian writers whose works dominated astrology throughout most of the last century threw up their hands in horror at such naughtiness and packed its perpetrators off to the fifth house. The seventh was far too dignified a place for them! The more straitlaced of their followers have tended to follow suit: seventh for formal relationships; fifth for the bit on the side. But this misses the essential point of discrimination between the two houses. As so often, we are required to think clearly about the distinction between an object and the function of that object.

To put this in a less mechanical way, we must never forget that even the cads and jezebels who tread the primrose path of dalliance are still people. As such, they are deserving of being given the seventh house. It is what we do with them that goes into the fifth: the eating out; the trips to the cinema; the sex. This is the dividing line: person seventh, activity fifth. This is true no matter how fleeting our acquaintance, and no matter whether or not we would admit to this relationship in front of our maiden aunt.

This lumping of all relationships into the seventh can cause confusion. As a practising horary astrologer, it would be easy to believe that no one ever gets involved with anyone who isn't already married to somebody else! In questions posed upon this theme, then, there are commonly two 'significant others', and they cannot both be shown by the ruler of the seventh. This is where the understanding of reception becomes so important, as it shows us exactly what the querent is thinking and feeling, and so enables us to identify whether it is lover or spouse who is shown by Lord 7, and then which planet shows the other of the two.

That said, we can often look straight to Saturn for the significator of the cheated spouse. As the Great Malefic, the Big Bad Wolf of the cosmos, he is an appropriate significator for the person the querent perceives as having no aim in life other than that of spoiling his fun.

For fun is one of the main meanings of this house: 'banquets, ale-houses and taverns', as William Lilly puts it. As such, it is a partner to its opposite house, the eleventh, which is where we find the friends with whom we share many of these joys. Not least of the pleasures which we are granted is that of sex – something forgotten by those who would consign it to the malefic eighth house. But the immediate function of sex is procreation, and it is in the fifth that we find the consequences of that: our children.

Again, we need to be precise in our thinking here. The fifth shows children and it shows pregnancy. It does not show pregnant women: these are shown by the same house that they usually occupy: my pregnant sister is still the third; my pregnant wife still the seventh. The astrology reflects our common-sense perception that they are still the same person, they just happen to be pregnant. Nor does the fifth show child-bed. As the old term 'confinement' suggests, this is a twelfth house matter.

The fifth shows our children in general; we have various means of deciding which of our children is which. For an enquiry such as 'Will I have children?' we can confine ourselves to the fifth. If, however, I wish to know which of my children has eaten all the cakes, or the answer to any question that requires us to distinguish one from another, we must find other significators.

We can take the fifth house and its ruler to show the oldest child, and the third house from the fifth (that is, the seventh) to show the next oldest, and so on, turning the chart three houses at a time. What we are effectively doing is taking the oldest as 'my child' and the next as 'my child's brother or sister'. Note that we take the third from the fifth, not the fifth from the fifth: this would show my child's child: my grandchild.

We can also look at the triplicity rulers of the sign on the fifth cusp to provide alternatives, as with brothers and sisters in the third house. This method uses the triplicity rulers given to us by Dorotheus of Sidon, which are:

Fire: Sun, Jupiter, Saturn
Earth: Venus, Moon, Saturn
Air: Saturn, Mercury, Jupiter
Water: Venus, Mars, Moon.

The first two rulers in each element reverse their order if it is a night-time chart.

The first ruler would show the oldest child (or the elder children out of many); the second ruler shows the middle one or ones; the third ruler shows the youngest. By weighing up their strengths, aspects and receptions we can then decide, for instance, which will have the greatest success, which would do best inheriting our business – or where the cakes have gone!

As a final option, we can look to whichever planets might be stationed in the fifth house. This is the least sound method; but we are sometimes confronted by questions where we need to find different significators for lots of children, and in such cases we cannot be too fussy about the methods we use.

In the past, one of the staple enquiries that kept the astrologer in corn was 'Am I pregnant?' Nowadays it is cheaper to pop to the chemist for an answer to that, but the astrological method is as valid as ever. The most convincing testimony is an aspect between either Lord 1 or the Moon and the ruler of the fifth house. Ideally, the planets would be in fertile signs (any of the water signs), or, at least, not in the barren signs (Gemini, Virgo and Leo). Finding any of these planets strongly dignified and in an angular house is another strong argument for pregnancy.

This is one of the very few questions to which we can find a favourable answer simply by the placement of a planet, without needing to find an aspect. So the ruler of the fifth placed just inside the first house – which gives us a perfect picture of the baby inside the mother's body – would give us a Yes. The Ascendant ruler inside the fifth house can show the same, but we do need to be more cautious here: it can show what the querent is thinking about or hoping for rather than what is actually the case.

These questions are usually quite simple: as the saying goes, you can't be a little bit pregnant, so the testimonies are usually unequivocal. The related question, 'Will I have children?' is more complex. Experience shows that this is tends to be asked only when hopes are already running low, most often when the querent is considering undertaking fertility treatment of some sort. As the chart reflects the life, borderline cases provide borderline charts.

A clear No is generally easy to spot: the main significators in barren signs or severely afflicted, especially by either Saturn or combustion. A clear Yes in these circumstances is rare. The best we can find is typically a statement that there is potential for fertility. That is, the testimonies are finely balanced, so treatment may tip the balance in the querent's favour. In questions of this type, it is important to include the seventh and eleventh houses in the judgment, the eleventh being the fifth from the seventh: it does take two, after all.

One of the easiest of all horary questions to judge is 'What sex is my baby?' Here we consider the gender of the various planets and signs. Aries and every alternate sign are male; Taurus and every alternate are female. Venus, the Moon and Mercury when occidental are female; the others male. We look to the sign on the Ascendant and the fifth cusp; the planets ruling those houses; the signs those planets are in. If in doubt, extend the search to the Moon, looking at its sign and the planet to which it next applies. There will be a weight of testimony in favour of either male or female.

It might seem that the concern of the fifth with 'pleasure, delight and merriment' would not play much of a part in the serious world of heart-felt questions with which the horary astrologer is presented. Yet such questions do get asked. Suppose, for instance, that you are planning a big party to be held in your garden. If you want to know what the weather will be like on the day in question, it is to the fifth house and its ruler that we would look. Or, if faced with the prospect of spending an arm and a leg on an evening at the theatre, it might be worth checking the fifth to find out if you will enjoy the play.

Other meanings of the fifth.

Just as the most common role of the eighth house in horary charts is not in its radical meaning as the house of death, but through its subsidiary significance as the second house from the seventh – other people's money – so the fifth is also important as the second house from the fourth. In days past, the astrologer would often be asked about the money that the querent might expect to receive from his father: how much is there, when will I get it, and will I have to battle with my brothers to lay my hands on it? Fortunately, such queries are rarer today.

Apart from the house of the father, the fourth is also the house of property, and the same significance as second from the fourth is relevant here. The fifth is whatever profit can be made from a piece of property. This is especially important if the property is being bought solely as a money-making venture: the state of the fifth and its ruler will then give us the 'bottom line' of the chart. But even if the place is being bought to live in, it is worth casting an eye over the fifth to gauge what its resale value might be.

The fifth is the house of messengers and ambassadors. This does not normally include the postman, whose job is to hand over a piece of paper about the contents of which he knows nothing: the meaning here carries a greater sense of involvement on the part of the emissary. The ambassador has a certain scope within which he can negotiate; the postman does not. This meaning takes us back to the spiritual framework of the chart, for the fifth is the house of the Holy Spirit, in whichever way this is expressed in the various revealed faiths. This is why Abu Ma'shar says the fifth is the house of guidance. Guidance, that is, from above.

Finally, the fifth bears an extreme importance in medical astrology. The great authority on this branch of the celestial art, Richard Saunders, who was a contemporary of Lilly, tells us that we must always refer to the fifth house in any medical query. This is because the fifth reveals the condition of the liver, and it is the liver that in the traditional medical model is the root of so many of our ailments.

As might be expected from the house of pleasure and delight, the associated planet here is Venus. Venus has her joy in this house, and not only that, but by the natural order of the planets from Saturn in the first and Jupiter in the second that gives the meanings of the houses, it is again Venus who is associated with the fifth. So she is doubly involved here.

On an elevated level, we are reminded that Venus is the planet that signifies the action of the Holy Spirit, as can be seen in the traditional iconography where the dove – the bird of Venus – descends as Jesus is baptised. On a more mundane level we are reminded that this is not the only house with which Venus is concerned. Following the same Chaldean order of the planets round the chart, we find that just as Venus is given the fifth, so she is given the twelfth. She may be all pleasantness, smiles and seduction in the fifth – but this same sweetness can lead us to our self-undoing if we are not on our guard.

THE SIXTH HOUSE

On our anti-clockwise tour of the chart, the sixth is the first of the malefic houses that we reach. Strictly speaking, this is something of a misnomer, as the house itself cannot do anything to harm us – unlike a malefic planet. But this is certainly not a nice place to be, and while the house may not be able to harm us, its ruler most certainly can.

The sixth is the house of illness. But illness is only part of the gamut of sorrows that it can provide. This is the house of the slings and arrows of outrageous fortune: of all the things that the harsh, cruel world and that odd bunch of people who inhabit it conspire to inflict upon us – always, of course, through no fault of our own. As such, it balances its opposite house, the twelfth, which shows all the dreadful things that we do to mess up our own lives.

It is the house of illness. Not, as it is often loosely called, the house of health. The person's health is shown by the first house, the house of their vital spirits. An illness can be seen as a battle for supremacy between the forces of the first house – the good guys – and the forces of the sixth – the baddies.

In the natal chart we read from the sixth the major ailments to which the native will be prone. We should not, of course, expect to see every minor ill shown here. Rather than a set of detailed medical notes, it is more like a fault-line running through the person. Everybody has such a fault-line, but in each of us it is different, causing our innate temperamental imbalances to manifest in different ways when they get out of kilter. These different ways are our particular illnesses.

The nature of the temperamental balance – whether there is an excess of fire, perhaps, or a shortage of water – will give a broad-brush guide to the way the system will react to periods of stress. The sixth house will help us to be more specific as to the presenting symptoms we are likely to find as a result of this imbalance. While it must be noted that traditional medicine does not share the modern obsession with sticking detailed labels on things ('You've got So-and-so's Syndrome'), it is nonetheless capable of specific diagnosis of great accuracy.

In the nativity, we would consider the sign on the sixth cusp, its ruler, and any planets that are stationed in the sixth, especially those close to the cusp. In a horary chart, we approach the illness differently. There, the sixth house tends to show where the illness is being felt, rather than either the true seat of the problem or its cause. For these, we must usually take the planet that is immediately afflicting our main significator. If, as is most often the case, the significator is in an uncongenial sign (cold/moist planet in a hot/dry sign, for instance) the prime suspect is the ruler of that sign.

It is not only ourselves or other people who get sick. Suppose we find the ruler of the second house in the sixth. How is our money? Sick. Similarly, in a question about house-purchase, finding the ruler of the fourth, which signifies the property, in the sixth is a serious warning to beware of structural defects.

While the sixth is our house of illness, it also has significance as the twelfth house from the seventh: the other person's house of secrets. So in any question involving seventh-house matters, be it a question of partnership, or a business matter, or a contest, we would need to be especially wary if this were where we found the other person's significator. This would be a strong indication that they are not being straight with us!

The house of work?

The other well-known meaning of the sixth is as the house of work; but this is quite wrong. We have dealt with this unfortunate product of Theosophist thinking in too much detail in *The Real Astrology* to need to rehearse the reasoning here. Suffice to say that the association of the sixth with work has no basis in any authority of standing. Nor is the sixth, as it is so often now called, 'the house of service', by which is meant the place that shows all the altruistic acts I do for others. It is not: it is the house of servants. I may not have an under-footman and a parlour-maid, but the man who comes to repair my TV or to plumb in my sink belongs here, as well as anyone whom I employ in my business. The idea that this house shows the services we do for others carries the odd connotation that doing something for someone else must be unpleasant.

So far as my plumber, or my employee, or the nanny who might look after my children is concerned, their job is a tenth-house matter. It is their career, their profession; so when examining their chart we would look to the tenth. If I wish to locate them in my chart, however – perhaps to find out why I always have problems with tradesmen, or if I am asking a horary 'Should I employ this cleaner?' – I would look to the sixth. There is no distinction between different qualities of job; it is a question of who is doing the work and who is paying the bill. After all, we all serve someone when we go to work, or there should be no reason for us to be paid.

Mercury, the planet associated with the sixth house, covers both of these main meanings. He is the natural ruler of servants. This is because he rules the reasoning mind, and the role of reason is to be a servant to the heart. This is something widely forgotten in the mercurial age that we inhabit, where the consequences of making reason our master are all too apparent.

Mercury is also, in one of his guises, Asclepius, the god of medicine. His is the caduceus, the serpent-twined staff that is even today the symbol of medicine. Another name for Mercury in this role is Ophiuchus. It is not inappropriate that the scientists (Mercury), with their over-valuation of reason, should be so determined to drag this so-called thirteenth sign into the zodiac.

Other meanings

The sixth is usually seen as a malefic house, so we would rather our planets were not located here, nor were in contact with the ruler of this house. Such placements or contacts usually show problems in whatever might be the relevant area of the life. The sixth does, however, have its happier side.

It is the home of small animals – 'animals smaller than a goat' in the traditional term – so this is where we find our pets. 'Smaller than a goat' is a generic term, so a Great Dane is found in this house, as dogs are smaller than goats, while even the tiniest of Shetland ponies would belong in the twelfth, which is the house of large animals and beasts of burden ('animals that are a mount to man' in the words of the Jewish astrologer, Abraham ben Ezra). In a horary chart for 'Where is my cat?' we would take the ruler of the sixth to signify the missing animal. With good fortune, we would find it applying to aspect the Ascendant ruler: she is on her way home!

As the third house from the fourth, the sixth shows my uncles and aunts (my father's brothers and sisters). If I were specifying uncles on my mother's side I might look to the twelfth, but generally we would go straight to the sixth.

There is another meaning of this house that causes confusion: that of tenants. In contrast to the idea of work, this does have authority in the texts; but we must be aware of a change in the meaning of the word. Nowadays a tenant exists on much the same level as his landlord: they are two equals who have reached a mutually acceptable arrangement on renting a property. In the past, things were different. If my tenant had the vote, for instance, I would expect him to obey my instructions on how he should cast it. There was an idea of deference with tenancy that is no longer pertinent today. This is why tenancy is given as a sixth house matter in the traditional texts – and why it no longer belongs there. This is not, it must be stressed, because astrology has changed; it is because the world that astrology mirrors has changed. Today, if I cast a horary on a question such as 'Should I let my house to this person?' I would take the prospective tenant as seventh house, in the same way that I would find a prospective buyer there if I were selling the house.

Even if we are faced with a decision like 'Should I sell my house, or should I rent it out?' which might seem to necessitate giving the seventh house to a buyer and the sixth to a tenant, we should avoid this – unless we really expect our tenant to labour on our land for a couple of days each week. In such an instance we would have to find a different approach, as both the buyer and the tenant would be shown by the seventh. The key is usually in the nature of the signs: fixed signs, implying permanence, would show a sale; cardinal or, especially, mutable would be more likely to show letting.

In medical astrology, apart from its general significance as the house of illness, the sixth has a specific connection with the intestines. If the significator of illness is placed in this house, that is exactly where we should expect the problem to manifest.

In mundane astrology it shows farmers and country-dwellers.

The planet that joys here is Mars. As always with the joys, there is a profund truth communicated by this fact. On an immediate level, the connection is straightforward enough: Mars is a malefic, and – with the exception of animals smaller than goats – we tend not to like the manifestations of the sixth. They hurt – just like Mars! But this Mars is a two-edged sword: it may be either turned against us, or wielded by us. The trick is to learn to pick it up and bear it in our own defence.

The malevolent interventions of the outside world in our life are justly symbolised by Mars. This is the sword turned against us. We are invited to take up a metaphorical sword and bear it against these problems: to take arms against a sea of troubles and by opposing end them. There are both external and internal sides to this.

On an external level, it is our duty to discriminate between what is right and what is wrong. Mars acts first like a razor, enabling us to divide the one from the other, then like a sword enabling us to put right what is amiss. The most obvious exercise of this is in medicine, where the application of Mars – the natural ruler of surgery – enables us by swift action to right the wrongs of the body. We can do the same in other areas of the life.

Looking more deeply, the sixth is, as it were, the house of alchemy. This too is ruled by Mars, as it is an art accomplished by fire, whether it is the external alchemy of bubbling retorts or the internal alchemy of the quest for spiritual perfection. This is, the traditional texts inform us, why we are given illness: to burn off the accumulated dross of our nature that we may arise into a better level of being. For this reason, the furnaces of the alchemists are typically shown as being built in humanoid form, making the point that what goes on outwardly in the furnace goes on also inwardly, and rather more importantly, in the heart, in a constant process of purification.

So next time you have a cold, rather than reaching for the various anodynes that modern medicine offers as a way of suppressing the symptoms, try embracing these various unpleasantnesses, seeing them as what traditional medicine suggests that they are: a way of ridding the system of a pernicious imbalance. Once that has been sweated out, we can then delight in the little rebirth that is our recovery, seeing it as a fresh start and an opportunity to do better in the new chapter that is beginning.

THE SEVENTH HOUSE

For most of us, the first introduction to astrology came as we attempted to find out if the girl or boy on the school bus was ever going to smile in our direction. Somehow the Sun-sign columns, by virtue of being written in black and white in the papers or magazines, carried more conviction than pulling the leaves off twigs to find out if she loved me or she loved me not. And as we started, so we continue, for in most astrological enquiries it it the seventh house, the house of relationships, that is the prime focus of interest.

The seventh is the house of the marriage partner, or the 'significant other' in our life. This is true even if the significant other in question is but a hopeless dream who persistently refuses to acknowledge our existence. So in a horary, the seventh is the house under consideration whether the question is 'Does the girl next door love me?' or 'Does Julia Roberts love me?'

There can be an ambivalence when we are considering whether to promote someone from eleventh house to seventh house duties – that is, whether to deepen a friendship. As ever, the thrust of the question will show the house concerned. So as the question boils down to 'Does she love me?' or 'Is there a future in this relationship?' it is usually a seventh-house matter, whether or not we already have some sort of relationship with the person involved.

It most certainly remains a seventh-house matter when considering the slightly less significant others: the passing fancies, or the bit on the side. As we saw when discussing the fifth house, it is 'seventh house for the person, fifth house for what you do with them'. This can, of course, create an ambivalence when there is more than one relationship on the go at once. In a question about a three-way relationship, the person actually asked about will usually get the seventh house. Receptions will guide us to the planet that signifies the other party. For instance, it is common to find either the planets of the two spouses to be ruled by the planet signifying the lover, or the planet of querent and lover to be ruled by the significator of the other spouse. The chart reflects the politics of the situation: the power that the players hold over each other.

Saturn is often significator of the other spouse, who is seen as the 'Great Malefic' bent on destroying everyone's fun. Any planet conjunct one of the main significators almost certainly shows an involvement elsewhere: in the old texts the word 'copulation' is used as a synonym for 'conjunction', which makes the point quite clearly enough!

A woman asked a horary: 'When will I meet a man?' Her planet was closely conjunct two other planets, so I asked her, 'What about the two men that you are already seeing?' 'They don't count!' she replied. Indeed, the lack of any reception between these conjunct planets made this clear: they may have come together, but there was no glue to hold them tight.

The opposition of first and seventh houses divides the chart into two. The eastern side, centred around the first house, is 'our' side; the western side, centred on the seventh, is for the other people. So if the natal chart has most planets clustered around the Ascendant, we usually find the native comparatively self-contained. If most planets are on the other side of the chart, the native has a keen involvement with the outside world.

When the Lord of the Ascendant or the Lord of the Seventh is placed near the opposite angle, the point is made all the more clearly. Lord of the Ascendant on the seventh cusp: you chase the boys. Lord of the seventh on the Ascendant: the boys chase you. A natal chart had the Lord of the seventh just on the first cusp, showing exactly that: the native was very popular with men. This planet, furthermore, was conjunct the Ascendant ruler by antiscion. As antiscion shows covert things, we see that she was forever being lured into illicit affairs.

In the nativity, we can learn a great deal about the marriage and how the partners will relate to each other by studying the first and seventh houses and the relationship between their ruling planets. We must always be aware, however, of the other house of marriage: the tenth. The further back we look in the astrological literature, the more we find marriage located in the tenth, rather than the seventh house. The reasons for this are still relevant today. The tenth is marriage seen as a social function: the dynastic or royal marriage would be an example here. The seventh shows marriage as a union of two people. Especially when considering natal or horary charts for people of Indian or Pakistani background, the tenth can have far more to do with the issue than the seventh.

There is a similar difference between the Arabian Part of Marriage and the Parts of Marriage of Men and of Marriage of Women. The latter relate to the tenth-house conception of marriage, while the former shows the seventh-house side of it.

Morinus called the triangle of houses based on the seventh the Triplicity of Marriage or Love. He said that there are three ways in which man is joined to his fellow man. Most importantly, 'is that of the body, which we call Matrimony' and which has the seventh house. Second in rank comes union by blood, 'which constitutes Brethren and Kindred, in the Third House'. Finally comes 'that of simple benevolence or favour, whence do arise friends, in the 11th house'.

Union by body is not, however, always harmonious: we are just as tightly engaged when we are wrestling – so the seventh is also the house of open enemies. The astrologer does not have so much scope for judging battle charts nowadays, but this meaning of the seventh is still important.

In judging charts for sporting contests, we weigh the first house against the seventh house to find out who will win. If the chart is a horary, we work mainly by the planets' relative dignity. If it is a chart cast for the start of the event, dignity has little importance and we work primarily from the movement of the planets. Such event charts can give results of great accuracy – even to the extent of allowing prediction of the final score, as I have demonstrated on TV on numerous occasions. The problem is that the techniques do not work for day-to-day matches. They work excellently for the high-profile, 'one-off' games like cup finals, but are no use at all at 3 o'clock on a Saturday afternoon, when fifty or sixty first-class football matches are kicking off at the same time, with similar charts. No help in winning the pools, then!

The first-seventh axis is also what we consider in charts for court cases. In a straight contest – an arm-wrestling match, for instance – we weigh the first and the seventh to see which is the stronger: that person will win. A trial chart has the notable difference that the case is decided not by wrestling one to one, but by the decision of a judge and jury.

In a horary chart about a civil action, the relative strength of the rulers of the first and seventh houses is less significant than their relationship with the rulers of the tenth and the fourth. If the Lord of the first or the seventh has a lot of essential dignity, it usually shows that this person has right on his side. Unfortunately, this does not necessarily mean that they will win.

The ruler of the tenth shows the judge, or, more broadly, the judicial system. It is not necessary to distinguish between judge and jury. If there is reception and aspect between the ruler of the tenth and one or other of the parties involved, that party will win – right or wrong.

The fourth house is 'the end of the matter', which phrase has a specific meaning in legal contexts: it is the verdict. In court, of course, everyone gets the verdict, whether they win or lose. In the chart, the verdict is like a prize: whoever gets to it will win. So if our horary shows the Ascendant ruler applying to conjunct the ruler of the fourth, we have good news: the querent gets to the prize and wins the case.

Criminal cases are different, as they are not a straight fight: they are the person against the Crown. When considering the outcome, we must look at what is going to happen to the accused's significator. If it is about to lose a lot of dignity, we see him being convicted. If improving its position, he will go free. These charts show with surprising frequency the significator entering its fall: the person is, quite literally, 'going down'. As a general principle, it is remarkable just how literally we can take the astrological chart.

As well as emotional partners, the seventh shows our business partners. Judgments on this matter need to be approached with a dose of common sense: it is easy to make judgments grounded firmly in fairyland. We should not expect to find the kind of relationship between business partners that we would between husband and wife. It is not even significant that there is no reception between their planets – provided that they both share an interest in the matter at hand. That is, translating the astrology into 'real life' terms: it doesn't matter whether or not they like each other; what matters is that they can work towards a common goal.

Many judgments go astray by feeding the wrong expectations into the chart. Do we want a business partner with whom we will enjoy working – as we make our way to the insolvency courts; or do we want a partner with whom we can succeed? If the latter is the case, his planet must show some essential dignity, to show that he has some gifts to bring to the business. Ideally, however, it would be strong, but not quite as strong as our own planet. We don't want to lose control!

The idea of partnership puts the doctor or the astrologer in the seventh house – if they are on the case at the time. Generally, doctors and astrologers, as people of learning, belong in the ninth. But if I am ill, my doctor becomes my seventh house in the chart for the illness. He is seen as my partner in the business of my getting well. Similarly, in a horary chart the astrologer is shown by the seventh house: the querent's partner in arriving at the truth.

If the astrologer asks his own question, he does not get the seventh as well as the first: one house is quite enough for anybody, astrologer or not. In charts on seventh house matters, meanwhile, we can assume that the seventh house has better things to do than signifying the astrologer, so we should not read ourselves into the chart.

While the seventh is the house of those closest to us, it is also the house of those of least significance. It is the house of 'any old person'. So if I ask out of idle curiosity, 'Will Madonna win an Oscar this year?' Madonna would be given the seventh house. It is said to be the house of fugitives, but I have always found more success in taking the natural house of the person concerned. So when my under-footman goes missing, I usually take him as the sixth house (servants) rather than the seventh.

A particular example of the open enemy is the thief. We may not know who he is, but the act of theft is considered as having made his enmity open, in contrast to the office gossip spreading scandal about us, who remains a secret enemy (twelfth house). If an article has been stolen the lord of the seventh is one candidate for significator of the thief.

We do need to be careful here, as thief and spouse are both shown by the seventh. Seeing the significator of the lost object in the seventh house can lead us to start slinging accusations of theft. Experience shows that, more often than not, it means only that the husband has picked it up and forgotten about it.

The planet associated with the seventh house is the Moon. On one level, the Moon is natural ruler of the people – of 'any old person'. But we must remember the basis on which our chart is constructed. The Ascendant is the place where the Sun rises. It is then, as it were, the natural place of the Sun. It is only fitting that we should then find the Moon, the celestial partner of the Sun, in the seventh, the house of marriage and of union.

THE EIGHTH HOUSE

Once upon a time, when someone had their horoscope cast the first thing the astrologer would determine would be the length of the person's life. This was considered an obvious preliminary, marking out the limits of the investigation. For there was no point whatever in the astrologer labouring to predict what would happen to the person on Wednesday if the chart showed that he was more than likely to die on Tuesday.

Things have changed, for predicting the length of life is now regarded as anathema. Many text-books of astrology state in no uncertain terms that it is one thing that no astrologer should ever do. Yet the awareness of its necessity still remains, for the astrologer will often be asked, 'OK: you've said such and such will happen; but what if I get hit by a bus tomorrow?' We may leave to one side the question of whether being hit by a bus is really as common a cause of fatality as such questioners seem to believe; but it is only reasonable to think that – making full allowance for the astrologer's human fallibility – if the client were to come to a sticky end in the immediate future, it could and should be read in the chart. Provided the astrologer is willing and able to do it.

We do hear appalling stories of irresponsible stargazers carelessly predicting ghastly ends with no apparent thought for the consequence of their words. Chiron and Pluto seem to be the favoured pegs on which they hang their dire prognostications, although neither of these has much connection with death in the chart. Obviously, we must be cautious about what is said and to whom we say it.

In more spiritual ages our ancestors knew that life makes sense only if lived in the full awareness of death: *momento mori*. To the modern eye, this appears gruesome. Far from it, as it heightens the capacity to live. Today, death is something that we much prefer to forget about, except when it happens to the bad guys on TV. Forget it as we may, it will still catch up with all of us, and it is an event of some consequence in the life.

It is often claimed that advances in medical knowledge (a somewhat loaded phrase, this!) and extended life-spans mean that death cannot be predicted by astrology. This is both demonstrably untrue and theoretically unsound. If such were the case, there must come a point in the life at which we are no longer subject to the stars. We might then all look forward to an old age where we are married to Mel Gibson or Nicole Kidman and win the lottery every week. Disappointing though it might be, this is unlikely to happen.

The debate about the prediction of death has created some bizarre ideas about the nature of the eighth house of the chart. As death has apparently been abolished, it has become necessary to find something else for the eighth to do. Favourite options today are 'transformative experiences', whatever they might be, and sex. The idea of sex as an eighth-house activity is quite horrific.

In any astrology that purports to say anything of concrete and verifiable accuracy, the eighth is the house of death. This is not death in any poetic or metaphorical sense, as some modern authorities claim. This is death in the very real sense of someone no longer being alive.

There are no prizes at all for predicting that somebody will eventually die: the important part of this prediction is getting the timing right. This is a matter of some importance, as if my astrologer has convinced me that I will die tonight, so I spend my last penny on an afternoon of hedonistic glee, I may not be best pleased when I wake tomorrow to find that the prediction was wrong.

Much nonsense is written about the timing of death from the chart. When Princess Diana died, for example, several published articles pinned the blame on progressed Pluto being in her eighth house. As at any one time a twelfth of the population has progressed Pluto in their eighth house, and they do not all drop dead, this has limited validity as a predictive technique.

We must indeed look at progressions, but we need tools far more precise than the infinitesimally slow meandering of progressed Pluto through the zodiac. Contacts with the eighth cusp and the ruler of the eighth house can be important, but even here we must exercise caution. Especially today, when we can flick up a progressed chart at the touch of a mouse, it is easy to forget that the progressions at any one moment are part of a system of on-going cycles, not a separate entity in themselves. So if we see the progressed Moon crossing our natal eighth cusp we should not panic, but remember that the progressed Moon circuits the chart every twenty-eight years, so in the average life-time it will do this two or even three times, usually without any ill effect.

The fixed stars assume a great significance whenever we consider the major turning points of the life, so progressions onto the more malign of them need to be considered. Malign, that is, from our own perspective, as the significance of the Moon's nodes makes clear.

At the end of his *Republic,* Plato gives a beautiful piece of astrological symbolism as he tackles the most fundamental issues of life and fate. He sees our life as a wheel revolving around a spindle. This is different to our common perception, which is of a straight line starting when we were born and moving inexorably to our death. The spindle around which our life is strung is the axis of the lunar nodes.

From our perception it is 'North Node good; South Node bad', as the North Node is the doorway into life, through which we come 'trailing clouds of glory'. The restrictive South Node, traditionally likened in its effects to Saturn, is the strait gate and narrow way through which we pass out of life. But, as the great teachers have ever told us, our perception is upside down, conditioned as it is by our viewpoint within life. It is significant in this context that we speak of the 'pearly gates' of heaven, as the lunar symbolism of the pearl brings us back to the Moon and her nodes, reminding us exactly of what we are talking.

From the progressions and the Solar and Lunar return charts the eighth house will show us the timing of death; it will also show us its quality: whether it will be sudden or long drawn; from illness or accident, or whatever.

As an extension from the idea of death, the eighth house also shows legacies. The major significators of wealth in the eighth house, William Lilly tells us, show 'profit from dead folks' – and he was well placed to know, with his happy knack of marrying rich widows shortly before they died!

The eighth is not money only from the dead, however, for as the second house from the seventh it shows the money of whomever the seventh house represents. If the seventh house is my wife, the eighth will show her money. This was a major topic of interest in Lilly's day, where much of the astrologer's practice was devoted to questions of 'How much money does my prospective spouse have – and how easily can I get my hands on it?'

This is not a matter that is entirely lost in the wastes of time. While wives now may not commonly bring dowries with them, we do often see in horary questions that there is a puzzling lack of reception between man and woman. 'What does he/she see in her/him?' we wonder. Until we notice that there is a strong reception between our querent and the ruler of the eighth house. 'Aha!' we think. 'He may not like her much, but he does like her money'. We can then consider the ruler of her eighth to see if she really has any money, or whether his interest is just wishful thinking.

As we saw with the second house, there is a deeper side to this. While the second is on a superficial level my possessions, and at this deeper level my self-esteem, so the eighth is the other person's possessions, and also his esteem for me. So often in horaries the eighth-house concern is less a desire for the other person's cash than the emotional necessity of their thinking well of me.

It is not only partners, but also opponents, and even 'any old person' that is shown by the seventh. So the eighth house is also my enemy's money. This is pertinent in horary questions of profit. 'Will I win by backing Red Rum in the 3.30?' What I want to see here is a nice aspect bringing the ruler of the eighth house – the bookie's money – to the ruler of either the first house (me) or the second house (my pocket). I hope to see the ruler of the eighth house strongly dignified and well placed in the chart: I want the bookie's money in the best possible condition. That is, I want a lot of it.

Its condition and the nature of the aspect will tell me how much I am likely to win. If there is a good aspect, but the ruler of the eighth is in poor condition, I may win, but I will not win much. This can help us make decisions. For instance, we may have the choice of a safe investment with a low return, or a higher return at greater risk. If the chart shows a big win, we may decide to take the risk; if a win but only a small one, we would take the safer option.

In a business context the other people are our customers, so 'the other people's money' is our takings. Again, we want to see the ruler of the eighth house in good condition – but there is an important rider here. If the ruler of the eighth house is in the eighth it will usually be very strong, as it will usually be in its own sign. But the ruler of the eighth in the eighth is a sure indication that, no matter how much money our customers might have, it is staying right in their pockets. This is all the more true if the planet is in a fixed sign.

On a more general level, the eighth is 'anyone else's money'. So, for instance, in vocational matters we commonly find the key significators in the eighth house when the person is an accountant, or in a similar profession whose dealings are with 'anyone else's money'. Mercury in the eighth house might almost be regarded as an astrological signature for accountancy, if Mercury in that chart has significance for the profession.

Through its associations with death, the eighth can also show 'fear and anguish of mind'. By this is not meant a specific fear, such as a phobia; but if in a horary chart the querent's significator is in the eighth house we would judge that he is seriously worried about the situation. As a general rule, the eighth is not a favourable place for a planet to be. Although it is a succedent house and succedent houses are stronger than cadent, it is, as it were, an honorary cadent house. Planets in the sixth, eighth and twelfth houses are significantly weakened.

In medical matters the eighth shows the organs of excretion. It is the opposite to the second house, which governs the throat, so the second shows what goes into the body, while the eighth shows what comes out. If there is a fixed sign on this cusp in a medical chart, and if the cusp is afflicted by either a planet in a fixed sign or by Saturn, which governs the body's retentive faculty, we might expect constipation, on either a physical or a psychic level. In a mutable sign and afflicted by a badly placed Jupiter, we might expect – again on either a physical or psychic level, as shown by other indicators – diarrhoea. The chart echoes the connection between money and eating and excreting that Dante shows in his *Inferno.*

The planet of the eighth house is Saturn. Following the Chaldean order of the planets around the chart from Saturn in the first, we come again to Saturn in the eighth. Saturn is the ruler of boundaries and of doors. As it shows us the doorway into life, it shows us also the doorway out. As it showed us the strait way in – we might remember what a hard passage is the birth – it shows us also the strait way out. From birth to death our life is bounded by Saturn – and our way out of these bounds is through Jupiter, the planet of faith, the builder of the rainbow bridge to the Divine.

THE NINTH HOUSE

Although the ninth might seem to be one of the less compelling houses of the chart – after all, most readings start and finish with the seventh house of relationships! – it can be seen as the key to the whole chart, especially if we wish to investigate the deeper levels of the psyche. This does not, it must be stressed, mean the unconscious: what is unconscious is unconscious for good reason, and is usually best left well alone. The ninth brings us face to face with the person's spiritual inclinations and capacities.

While the reaction to this may too often be 'So what?', it is at this level that the natal reading has its true purpose. For as astrology makes sense only in its relationship to the Divine, so our lives make sense only within a spiritual background. This tenet was taken for granted throughout most of astrology's long history.

The usual model of the natal reading by traditional method follows the analogy of building a house: we start from the bottom up, laying the foundations before we tackle tasks such as putting on the roof. So our first job is to assess the native's temperament. This gives us a broad judgment of type, against which all the more detailed parts of the reading must be seen.

For all that the modern astrological 'cook-books' tell us about the aspects in the natal chart, these aspects make sense only when seen against the background of the temperament. We must always ask ourselves whether the person is choleric, melancholic, sanguine or phlegmatic (fiery, earthy, airy or watery) by nature if we are to know how the particular aspect will operate. Once we have laid the foundation that is the temperament, we will find that the rest of the reading falls into place quite easily, without many of the contradictions which the modern concept of 'synthesis' attempts to resolve.

There is, however, another, deeper, model of practice through which we can approach the chart. Here we begin not with the material foundations of the nature, but with its highest levels: the spiritual possibilities. This might seem like a contradiction to our original analogy, presenting us with the feat of putting on the roof before the walls are in place; but it should be seen not so much as starting from the top and working downwards, as starting from the inside – the veritable core, or essence, of the person – and working from there to the superficial traits of character.

To examine the native on this level we look to the ninth house of the chart and its relationship to the third house, which is where the potentials of the ninth are or are not put into practice. Also of the utmost importance are the seven key Arabian Parts: these can indeed be seen as the Seven Pillars of Wisdom on which our judgment is founded. The numerical echo is no coincidence.

These seven Parts are those based on the Part of Fortune. They are Fortuna herself (Asc + Moon – Sun); the Part of Spirit (Fortuna reversed: Asc + Sun – Moon); the Part of Love (Asc + Part of Spirit – Fortuna); the Part of Despair (Asc + Fortuna – Spirit); the Part of Captivity and Escape (Asc + Fortuna – Saturn); the Part of Victory and Aid (Asc + Jupiter – Spirit); the Part of Courage (Asc + Fortuna – Mars). The formulae for these Parts are often reversed in night-time charts, but the reasoning for doing so is questionable, while keeping the formulae unchanged works well, whether in plumbing the depths of the nature or in such mundane, but easily verifiable, matters as forecasting the football scores.

The names that these Parts have acquired in English understate their true significance. The Part of Love concerns Love on a far wider scale than our usual preoccupation with romance – although this is indeed part of it. It encompasses all our human bonds, and stretches farther yet to tell much about our aspirations and motivations. The Part of Despair is the unfortunate awareness of the limitations of our human condition, and where in the life this pinches us the most.

The Part of Captivity has its obvious connections with prisons; but most of us go through our lives without any connection with prison. It shows the unnecessary bonds we put upon ourselves by our own foolishness (in contrast to the inevitable bonds that are shown by the Part of Despair). As the planet connected with this Part is Saturn, god of doors, it shows us not only the way into these prisons, but also our way out. On the broadest level, the imprisonment of which it speaks is that of the soul within the world: our own incarnation.

The Part of Victory has little to do with winning at dominoes, but is a reminder – as its planet, Jupiter, makes clear – that victory is ever a gift from above. It shows us aid from the Divine, the 'daily bread' that sustains us both physically and spiritually. Finally, the Part of Courage shows greatness of soul, of which physical courage can be a part.

When assessing the ninth house's role in this we must look to the house itself, its ruler and any planets in that house, especially those close to the cusp. The modality (cardinal, fixed or mutable) of the sign on the cusp and the sign in which the house ruler falls will tell us something of how constant in faith the native might be – remembering, as ever, that this must be seen within the context of his or her whole nature. It is always worth having a look at the condition of Jupiter too, as he is the natural ruler of faith.

If we find the house ruler in its detriment and the house itself afflicted, things do not look good. This does not mean that the person should be written off as the worst of sinners: such crosses can be the spur to work hard at the spiritual life and achieve greatly. The nature of the planet afflicting the house will tell where the major problems lie.

Suppose it is Saturn in Aries, poised on the cusp of the ninth. The reason Saturn is so weak, and therefore so malign, in Aries is that the speed of that sign is incompatible with Saturn's plodding nature: this often manifests as impetuosity. So we might judge that the native's desire for instant solutions makes any sustained spiritual effort that much the harder.

Or we might link the afflicting planet back to the house it rules. A typical example might be the Lord of the tenth house afflicting the ninth: the native's all-consuming commitment to his career leaves no time or effort for higher things. Similarly, we can trace the significance of a helpful planet in the ninth through the house it rules. Maybe I have the ruler of the eleventh house casting a fortunate ray over the ninth: I have friends whose company guides me to a higher path.

It is from these spiritual concerns that the ninth draws its lesser meaning as the house of travel. This governs long journeys; but what is a long journey? The ancient texts define it as a journey where we get there and back in a day, or maybe two. Shorter journeys belong in the third house. But the real dividing line is whether the journeys are routine or special. The trivial journeys that make up our daily round are third house; any special journey, no matter how near or far, echoes the nature of the pilgrimage to God that is the journey of our life, and so belongs in the ninth.

Indeed, the Arab NeoPlatonists, the Brothers of Purity, saw the ninth as the image of the whole of the life, finding significance in the fact that if we give one month per house from conception, which we may take as the Ascendant, the ninth cusp will show the birth, the start of our voyage.

In most of our enquiries we are concerned more with mundane journeys, and the ninth will tell if the native has an inclination to travel, and how successful such travels will be. If we are looking for the success of one specific journey in a horary chart, we would also check the condition of the tenth house. As the second from the ninth, this shows the journey's money – the profit that will be made. A strong Jupiter on that cusp should have us packing our bags with enthusiasm!

So also the tenth shows the profit from another ninth house concern: our studies, or our knowledge. The condition of the ninth house in a horary on this subject will show if we know anything; this is not necessarily connected with our ability to make money out of it. For this, we look to the tenth. As with questions of faith, so here the failings in our knowledge will be revealed by afflictions to the house and its ruler.

Knowledge is seen as a kind of journey. Another journey is a dream. Dreams in the loose sense of 'I dream about dating Miss. Wonderful' are eleventh house issues: ambitions. In the strict sense, as in what we experience whilst asleep, they belong here (not in the twelfth). The dream is seen as a source of information, which may or may not be correct. In the classical image, the dream comes through the gate either of horn or of ivory. Those which come through the gate of horn are true, as if horn is worked very thin it becomes transparent, while ivory remains opaque no matter how thin it might be.

In a horary chart, the condition of the ninth will tell us if the dream is to be believed. Or, for that matter, any other prophecy or prediction: 'my psychic told me...' If we wish to use the chart to interpret a dream, we are best to regard the dream as if it were as real as anything else in our life. Thus we ascribe the houses exactly as we would normally do: I am first, my wife is seventh, my boss is tenth, and so on. By analysing the chart according to the usual rules we can find a concrete meaning for the dream, avoiding the psycho-jargon with which dream interpretation is so often plagued.

So, for instance, if I dream I am arguing with my boss, by unravelling our motives in the same way as we would do in any other horary chart, by studying the receptions of the two significators, we see the reasons for the argument, while the house placements will allow us to reveal any deeper layers of meaning in the dream's choice of characters.

The ninth is often prominent in horaries for 'When and where will I meet my husband?' questions. If the planets of querent and potential spouse come together in the tenth, we see them meeting at work. Unfortunately for the astrologer, most other options for meeting are covered by the ninth: evening-class, on holiday, in church. If this is the house emphasised, it can be hard to decide which will be the one. For Asian querents, the ninth has a specific meaning here: that of the marriage bureau. As the bureau has taken the role of the wise man who would arrange a match in the past, so it takes the same house of the chart. The same cannot be said of the western agencies.

Following the Chaldean order of the planets that has explained why each house means what it does, we find Jupiter as the planet associated with the ninth house. While it is not usually a good idea to mix eastern and western schools of astrology (no disparagement to the Vedic methods, which are fine indeed: they just don't mix successfully), the Indian name for Jupiter, *Guru,* makes the point perfectly. Jupiter is the builder of the rainbow bridge from the human to the Divine, and this part of the chart is from where that bridge leads.

This point is repeated by the Sun having its joy in this house, as the Sun is the visible symbol of God within the cosmos. This is an idea which the historians and archaeologists are prone to mistake, so it is worth clarifying exactly what it means. No civilization has ever worshipped the Sun (or the Moon, for that matter). What various civilizations have done is to take the Sun as a naturally occurring symbol, not of God, Who is beyond all symbol, but of His action and manifestation within Creation.

Once, the traditional sources tell us, the house of God that is the ninth would have been on the Midheaven, its rightful place. But when Man fell, everything slipped out of kilter. This is reflected by the planets' moving in elliptical, rather than circular orbits. The effect of this transposition on the mundane houses is that they all shifted round by one place. The point made by this piece of cosmology is that with the Fall, there came into Creation two differing viewpoints: God's and Man's. Before then, when Adam would walk with God in the cool of the evening, the viewpoints were as one.

THE TENTH HOUSE

The Midheaven, the cusp of the tenth house, dominates the chart, and this dominance is reflected in its significance in the natal reading. For while the tenth is commonly associated with the work that we do, this has an importance far wider than the means by which we keep food on our table.

Ptolemy, who is, for all that he is neglected today, by far the most influential of all writers on astrology, speaks of the tenth house in a section devoted to 'action'. The main role of what he terms 'the lord of action' is to show what profession the native might practice; but it has much to do with the quality of any action that we might undertake, involving not only our work, but the manner in which we do anything. It helps show the 'how' of our behaviour. If the first house shows the nature of this particular incarnation, this soul in this body, the Midheaven shows this incarnation in action – the soul going for a walk. Our judgments here amplify and refine the basic details of the 'how-ness' of the person that we have found from the assessment of temperament and manner which forms the first stage of traditional natal method.

Ptolemy's judgement of this part of the life begins with the Sun. Not, it must be emphasised, with the Sun itself, and so not with the facile numbering off of all Taureans to work on farms and all Capricorns to be business executives, but with the planet that is closest to the Sun at sunrise on the day in question. Preference would be given to an oriental planet (one that rises before the Sun). Oriental planets find it easier to shed their influence on the world. This reflects the astronomical reality: an oriental planet rises into a dark sky, so it can be seen clearly. When an occidental planet rises the Sun is already up and so it cannot usually be seen at all.

We must combine our assessment of the planet rising closest to the Sun with that of any planet situated near the Midheaven, especially if it is in aspect to the Moon. The lunar aspect helps to, as it were, vivify the planet, almost as if this aspect were what plugs the planet into the mains and so allows it to work. This empowering of other planets is the main role of the luminaries, to which role the manifestation of their own natures is very much a secondary consideration.

The Sun is seen as the ultimate source (within the cosmos) of all power, while the Moon, the closest planet to our Earth, acts as a kind of lens through which all this power and all planetary influence is filtered on its way into what the ancients termed 'the world of generation and corruption' – the world which we inhabit.

The significators thus identified are considered together, with emphasis being given to whichever of them is the stronger by essential and accidental dignity. In many charts we will have only one such planet, and in some none at all, for it is only planets that are close-ish to the Sun that can be considered under our first criterion. As always, they are judged together with the signs in which they fall, the signs being the adjectives to the nouns that are the planets.

If our first criteria do not produce a planet, we then – and only then – turn to the ruler of the tenth house. This, Ptolemy warns us, will tend to show 'the occasional pursuits of the subject, for persons with such genitures are for the most part inactive'. Like all such sweeping statements in the texts, this is 'all things being equal', and there will often be other indications which will counterbalance it.

The planet or planets and their signs will combine to tell us about the profession. If other considerations of temperament and manner were in accordance, for example, we might decide that Mars shows that our native is a soldier. Suppose Mars is in a water sign. What sort of soldier? A wet sort of soldier: so a marine, or a sailor on a warship. Perhaps this Mars is in close trine aspect to a strongly dignified Jupiter: we might further decide that this must be a very successful wet sort of soldier – a fighting admiral, perhaps. But maybe Mercury is combined in judgment with this Mars, and Mercury is by far the stronger of the two: we might have someone who writes about wet sorts of soldiers, or perhaps devises tactics for their actions.

Mars, Venus and Mercury have an important general role in the assessment of *magistery*, which is the traditional term equivalent to profession or vocation. They denote, broadly speaking, whether we work with our brain, our brawn, or our aesthetic senses. Any one of these strongly placed in either the first, seventh or tenth house will have a major influence on the profession. Indeed, for want of any more commanding testimonies, we can often plump for whichever of these three planets is the stronger and hang our judgment on that.

It is important to follow these stages of judgment rather than jumping straight to the ruler of the Midheaven: to do that can be misleading. My experience suggests that what this planet often shows is what the native can do – a kind of fail-safe or default option, into which they will fall whenever they are not pushing themselves to do something more interesting. When lives reach crisis point and the native decides that he can no longer cope with a certain line of work, the chart often shows him abandoning the ruler of the tenth in favour of the strongest vocational indicator, which will show a job more emotionally nurturing.

There are two Arabian Parts of particular relevance here. The Part of Vocation is the same as the Part of Fortune, except that it is extended from the Midheaven rather than from the Ascendant. As the Part of Fortune shows the person's 'treasure', or what is their deepest concern, so the Part of Vocation shows their treasure in working terms, or in terms of action. The Part of Fame, or of Work to Be Done, can show that for which a person is most noted. It is often illuminating in the charts of people famous in their field. The formula is Ascendant + Jupiter – Sun by day, or + Sun – Jupiter by night. As the name 'Work to Be Done' suggests, it carries a sense of duty: this is where 'a man's gotta do what a man's gotta do'.

The connection of the Midheaven with career, or with the quality of action, has a special significance when we must judge charts for which we have no time of birth. The convention is that we use a chart set for either noon or sunrise on the day in question. With a sunrise chart the Sun will be on the Ascendant, weighting that part of the chart. These charts are best if we wish to explore the native's private life. Noon charts, with the Sun's position emphasising the Midheaven, are much the better if we wish to investigate the native's public life or career. With the charts of celebrities, it might be suggested that the latter is the only fit cause for investigation – the private life being no more the just concern of the astrologer than it is of the paparazzo.

While the tenth house has significance for our job, it also shows our boss. The tenth is the house of kings, being the proudest of houses, as the place which the Sun holds at midday. By analogy, it also shows the ruler in any specific situation. This can cause confusion in horary questions on career issues. If the tenth house shows the job, where do we find the boss, if we need to distinguish between the two? We can sometimes turn the chart to take the tenth house from the tenth (the seventh). So if the tenth is the job, the tenth from the tenth is the boss of the job. If the question is 'Will I get the job?', however, this house is already in use, as the seventh is the house of our open enemies, who are in this context our rivals for the position. In this case, we can take the planet that disposits the tenth ruler. Lord 10 is the job; its dispositor is, literally, the ruler of the job: the boss.

In court-case questions the tenth shows a specific kind of boss: the judge. Although the texts refer to the tenth as 'the judge' it can nowadays be taken as showing the whole court system. It is not usually necessary to distinguish between the judge and the jury by finding separate significators for them: this does nothing but add an unnecessary complication. The tenth ruler has immense importance in these charts, for it does not so much matter who is right and who is wrong, as it matters whom the court prefers. A strong contact between the significator of either party in the case and that of the judge will override any other consideration in the determination of the chart.

As the natural position of the midday Sun, the tenth is the house of glory and of honour, so success is found here. It is common, for instance, to look to the ninth house when considering questions about examination results. But the ninth house shows our knowledge: there is not necessarily a direct connection between this and our passing or failing the exam. What we are really asking is 'Will I succeed?' which is a tenth house issue.

It can be difficult to decide whether or not to turn the chart for certain tenth-house matters. As a general rule, the bigger the thing, the less likely we are to turn. Extreme cases are clear-cut. If I ask 'What are my son's career prospects?' the career belongs specifically to him, so we would turn the chart, taking the tenth house from the fifth. If I ask 'Will my son win Olympic gold?' this event is so public and of such note that we would look straight to the radical tenth. There is only a fine line of division between some less extreme cases. 'Will my son get a holiday job in the supermarket?' is small-scale, so we would turn. 'Will my son be the next manager of Manchester United?' would take us to the radical tenth. Somewhere between the two there is a dividing-line; if we are in doubt, close study of the chart will always reveal on which side any particular instance falls.

As discussed in the article on the seventh house, the tenth has a lot to do with marriage. The further back in the texts we look, the more marriage reverts from the seventh to the tenth house, being seen less as a union of two equals and more as a social institution This has significance today when considering arranged or introduced marriages, in contrast to the 'we're in love, let's get married' of the seventh house.

Although there are many now who choose to disagree, several thousand years of astrological tradition – and the small matter of demonstrable accuracy – places the mother in the tenth. This is an extrapolation from the fourth house, which shows our parents (our root: the base of the chart) in general and our father in particular. Our father's wife, then, must be shown by the seventh house from the fourth: the tenth. Whether this fits with our conceptions of social justice is neither here nor there: the heavens have not yet started to arrange themselves to suit our ephemeral fashions of thought.

On a simple and graphic level, the tenth has significance through its position at the top of the chart. If, for instance, the significator of a lost object is found in this house we might judge that the object is high up, or in the attic or loft. Similarly, if its significator were at the bottom of the chart, in the fourth house, we might judge that it was in the cellar.

That the tenth is so high explains the ambivalence about this house in the old texts. This was not seen as a happy house; for, following the model of the ever-revolving Wheel of Fortune, once you have reached the top there is only one place that you can go! This is also the thought behind the reputation of the royal star Regulus – strongly associated with the Midheaven – for giving glory and then taking it away again. This does not necessarily mean that there will be a fall from office: the unavoidable fact that even the mightiest of emperors must yet die, just as the Sun – the planet of which Regulus is the likeness – must always set, is sufficient explanation.

No planet joys in the tenth house. The Chaldean order gives Mars as the planet associated with the tenth. Mars has its obvious associations with kingship and empire, as the strong right arm carves out its area of rule. But no matter how mean or trivial our job might be, the connection with Mars is none the less real. Our profession, or our magistery – 'that of which we are master' – may not extend across the globe, but even if it is nothing more than a yard of production-line or a check-out in a supermarket, still it is our own little empire, conquered by our ability to fill its necessary functions.

THE ELEVENTH HOUSE

As if it were a many-faceted jewel, we can see the symbolism of the structure of the chart from many different angles, each of which carries its own distinct, but absolutely congruous, picture of the truth. From whichever direction we approach the eleventh house we find one point repeated: this is the happiest of the houses. Not, that is, the strongest; for strength and happiness do not necessarily go together – but undoubtedly the most fortunate. The Midheaven, for example, is a much stronger house. A planet poised on the tenth cusp will dominate any chart, and is ideally placed to turn this dominance into decisive action. Yet the tenth, for all its power, is not seen as a happy place, for the Wheel of Fortune, deriving from the Sun's primary motion around the Earth – its passage from dawn to high noon to sunset – shows that having attained the pinnacle there is only one place for a planet to go: downwards. Uneasy lies the head that wears a crown.

The eleventh is seen as the most fortunate house as it is on the way up, and stands tip-toe on the threshold of success. It is for this reason, the Roman writer Manilius explains, that the eleventh is particularly associated with Jupiter, most fortunate of the planets. It is in the eleventh that Jupiter has his joy.

The concept of 'joy' seems to cause some confusion: it must be distinguished from the idea of rulership. That Jupiter joys in the eleventh does not mean that he rules that house. It becomes clearer if we think of the chart as a little village of twelve houses. We can see that Jupiter rules one of these houses: being Jupiter, this would be the big house with the long drive and dozens of servants. There is, however, another house in the village. It is only a small place, without rich array. But the people who live there are so good-hearted and optimistic that this house is filled with warmth. After a hard day organising his many servants, there is nothing Jupiter likes more than to drop round and relax in this pleasant atmosphere. Similarly, Saturn finds the austere and gloomy conversation of the house next door – the twelfth – to his taste, so he likes being there. This is the essence of planetary 'joy': it is, as it were, a part of the chart where that planet likes to hang out.

We can arrive at Jupiter's association with the eleventh from another direction. As our astrology is – and must be – congruent with and dependent upon revealed faith, we can trace the spiritual history of mankind in the chart. The Ascendant gives us our starting point, the place of incarnation: our creation. The twelfth is our self-undoing: the Fall. Hence its being the first of the 'cadent' or falling houses.

To get from first house to the twelfth we must go in the opposite direction to the order of the houses. To go in this direction, we follow the primary motion of the Sun. So doing, we bind ourselves to Time, which is at once the reason for and the consequence of our Fall. This is an example of the astrological truths that are to be found in the traditional writings on the seven days of Creation. As so often, we learn far more about astrology from books that are not overtly about the subject!

Man having fallen, the next great milestone is the Flood and the covenant that God then makes with Noah, as shown by the rainbow. This is the eleventh house, and the story is heavily Jupiterian.

Jupiter is the god of rain: one of his altars in Rome was dedicated to Iuppiter Pluvius – Jupiter the rain bringer. This is Jupiter as lord of fertility, for, of course, no rain, no growth. Jupiter never being one to do anything by halves, when he brings rain he brings big rain. The astrological texts on weather forecasting refer directly to 'the opening of the gates of heaven' before the Flood in their testimonies on heavy rain.

More than that, however, for Jupiter is also associated with the rainbow. The imagery here is perhaps most clearly seen in the picture of the rainbow bridge that leads to Valhalla in Norse myth, but the concept is ever the same: Jupiter is the builder of the bridge from human to Divine. Most particularly, the rainbow, which is the reminder of God's covenant, is that most fortunate of all things: the promise that there is redemption after the Fall.

The treatment of water in the tradition is somewhat different from what we might expect today. Water is seen less as the pure and vital stuff we get from a tap or bottle than as the tempestuous destructive stuff that obstructs the passage from land to land – hence the phlegmatic temperament, with its strong and unfocussed desire nature (shown by Mars being ruler of the water triplicity) being traditionally regarded as the most difficult of the four temperaments. The role of Jupiter is crucial, for it is what might be called 'the Jupiter cycle' that takes this destructive sea water up into the clouds from where it falls as the potable sweet water on which we depend for life. So also the other 'Jupiter cycle', whereby the praise which we render to the Divine swells the refulgence in the courts of Heaven, bringing increased blessings showering down to sustain us.

So it is that the eleventh is the house of friendship. We English are very free with our use of the word 'friend'; traditionally – and in other cultures – it is a treasure more closely guarded. The distinction in some languages between, for instance, *tu* and *vous* goes part of the way to showing the difference. But at its root, my friend is someone who helps me out of good-feeling towards me. It is not just someone with whom I share a cup of tea and a chat from time to time. Least of all is it someone whose company I enjoy but who leads me into bad habits.

Friendship, then, is a mundane reflection of this covenant of redemption after fall. Indeed, we might on a more immediate level see the capacity to forgive as one of the marks of a true friendship. This is friendship as Cicero's 'one good thing': the channel of benevolence into my life.

A particular benevolence that many of us dream of being directed into our life is a handsome win on the lottery, so as another example of this same underlying idea such unmerited winnings are shown by the eleventh house. If I win no matter how huge a sum by outwitting the local bookmaker, the money I win is shown by the eighth house: my enemy's money. It is large amounts won by chance – the lottery, premium bonds or football pools – that fall into the eleventh.

We see three different levels of profit here, for the eleventh is the second house from the tenth, and as such shows the King's money and my boss's money. In itself, as bounty dropping from the skies, as an example of the beneficence and compassion, it is 'pennies from Heaven'. It is also what is traditionally termed 'the gift of the King': favour from high places. And, as the boss's money, it is my wages.

All of these carry the theme of a lack of desert. We may share the writer's view that we are deeply deserving of a lottery jackpot, though we might find it hard to explain quite why. The gift of the King, as it is granted only from his mercy, must of necessity exceed our deserving. So, in the astrological view, must our wages. That we have them at all, that our farm does not blow to dust, is cause to be thankful – which returns us to Jupiter.

It is no coincidence that Jupiter and Mercury between them rule all four of the double-bodied signs. For, as Ibn 'Arabi explains, these signs are those that show the return to the source from which we came, and our means of doing this is prayer, praise and gratitude: all of which, in their different ways and different levels, are shown by Jupiter and Mercury.

Although it may seem to have a connection with the idea of friendship, the common attribution of 'social organisations' to the eleventh house must be refuted. This is one of the many errors that have been introduced to astrology through the simplistic formulae of the Alphabetical Zodiac, the theory here being that the eleventh house equates with Aquarius which equates with Uranus.

Leaving aside the facts that there is no rational connection of Uranus with Aquarius, or of Uranus with friendship, we need but point out that social organisations are not relevantly connected with the concept of friendship as discussed above. Society, the undifferentiated others among whom I dwell and with whom I must deal, is shown by the seventh house; a club, where I may meet my chums, by the fifth. So if I cast a chart about the merits of joining, say, a trades union, the eleventh is not the place to look. If, however, my horary is about the wisdom of taking a job, the eleventh will be extremely pertinent, as any affliction there is likely to have an adverse effect upon my pocket.

If assessing the quality of friendships, whether it be a particular friendship in a horary or the general nature of the native's friendships in a birthchart, we would look at the ruler of the eleventh house and whatever planets are either in that house or casting their aspect closely to its cusp. We would like to see indications of stability, shown by fixed signs, indicating that friendships will last; the relevant planets in their essential dignities, giving friends of inner quality; and a good degree of mutual reception with the ruler of Ascendant, showing that the native will get along with them. For I may, in principle, have the best friends in the world, but if I treat them with disdain, or am too disordered myself to appreciate their qualities, they will render me but little good.

On a more abstract level the eleventh covers such things as trust, hope and confidence. The eleventh is 'the house of hopes and wishes'. This is sometimes pertinent to judgment; mostly it is not. Too often student astrologers will try to find a happy answer to a horary by connecting planets to the lord of the eleventh as a way of granting the wish; but this does not work. That someone has bothered to ask a horary shows that there is a wish – yet we cannot judge all questions by reference to the eleventh house. If the seventh house shows that Susie cannot stand me, it is no good my looking to the eleventh in the hope that this might change her mind.

The text-books do use this idea of the underlying wish behind any horary question by giving the eleventh as the house to consider if the querent refuses to divulge the nature of his question. The suggestion is that the question, whatever it might be, is treated as 'Shall I obtain my wish?' Any client who wanted to ask a question but refused to say what it was would get a very short answer from at least one astrologer.

As the second house from the tenth it represents the king's advisors or aides. So if the Prime Minister is shown by the tenth house, his cabinet is shown by the eleventh. There is a clear distinction between his advisors and his employees. The cabinet is second from the tenth; the person who sweeps up after its meetings is sixth from the tenth: the servant of the king.

In medical charts the eleventh shows the lower portion of the legs, down to the ankles. We must exercise a little common sense here. With the legs covering the whole chart from the tenth, or even the ninth, house down to the twelfth there is a great deal of space and not a lot of action. This is most of all apparent in the eleventh, if we take the texts too literally. If we restrict this house to the shins, we will never find it of use, unless we deal regularly with footballers or hockey players.

The legs, as indeed the whole body, must be seen as a gradual development through the signs, not as a series of disparate chunks. So as the thighs start at the cusp of the tenth, we can run through the house until we find the knees at its very end, or on the cusp of the eleventh. This satisfies both reason and experience.

Following the Chaldean order of the planets, it is the Sun that is associated with the eleventh. As is only fitting, for as the eleventh shows the good things that descend to us from Heaven, so the Sun is the image of this endless, inexhaustible bounty permeating and sustaining the cosmos. The Sun is in traditional terminology 'the Lord of Life'. It is on the Sun, both directly and, as its rulership of staple foods, such as cereals, makes clear, indirectly we rely. So, as shown for example in the prayer 'give us this day our daily bread', the concepts of Sun and Jupiter are reconciled in the significance of this most fortunate of houses.

THE TWELFTH HOUSE

Morinus gives a description of the last of the houses, the twelfth, that makes it sound like the setting for a new episode of the adventures of Indiana Jones. It is, he writes, 'the valley of miseries, which contains all the miseries of this Life, and also the House of the secret enemies, of the World, the Flesh and the Devil'. While the sixth and eighth houses might justly feel somewhat hard done by at being deprived of their fair share of 'the miseries of this Life', Morinus' view that the twelfth is not a nice place is clear, and is repeated as strongly if perhaps less colourfully in the writings of all the traditional authorities.

Morinus explains exactly why the twelfth is 'the dark den of sorrow and horror'. If we place the native at the point of birth in the chart – if, that is, we put him right on the Ascendant – he has a choice of two directions in which he may travel. He can follow the direction of the houses, travelling anti-clockwise around the chart, or he can follow the direction of the Sun's primary motion, travelling from the eastern angle up towards the Midheaven. The first of these, he says, is the way of 'the Rational Appetite, whereby a man is carried by the Motion of the Planets in the way of Descension and humility', as shown by the downward motion thus involved. The other way is born of pride, of man's unnatural aspiring to the height of the Midheaven, and is therefore the way of 'the Sensitive Appetite' – by which he means of course something rather baser than the ability to empathise with one's fellow man.

The connection with the story of the Fall is obvious, as overweening pride leads Man into the house of self-undoing. It is significant that this direction is that of the Sun's primary motion, which is our immediate indicator of the passage of time, the consequence of the Fall being our bondage to the illusion of time, our days being numbered. We see here the importance of the association of the twelfth with 'the hatred, deceits, Machinations, Treacherousness and Injuries of Enemies, especially Secret ones', for at the deepest level our secret enemy is the Tempter – the tempter who, for this is the house of self-undoing, resides within ourselves.

On a more immediately practical level, the twelfth-house enemy is usually the unnamed and unknown person who plots against us. Even a thief, by virtue of the openness of his action, is found in the seventh house of open enemies. The secret enemy is the unknown tongue who spreads slanders in the office, or who takes strange delight in unpinning our best laid plans.

The one named enemy who is found in the twelfth is the witch. While witchcraft might seem to belong only to some bygone age, horary questions on this subject are by no means rare, and by no means always from far-off cultures with different assumptions from our own. Sometimes these questions will be phrased in terms of witchcraft; sometimes they come in fashionable disguise: 'psychic attack' is the common modern phrase for the phenomenon. It must be made clear that the witch belongs in the twelfth whether claiming to be white or black. As demonstrated by the unfortunate psychic knots into which horary clients have so often tied themselves through either dabbling in white witchcraft or associating too closely with those who dabble, the apparent distinction in colour is largely illusory.

In his chapter on twelfth house matters William Lilly provides a handy recipe for dealing with witchcraft. The operator must follow the witch home, and then, before anyone else has entered the house, pull a handful of thatch or a tile from above the door. This is then boiled up with the victim's urine. This is one of the Master's techniques that I have not found it necessary to test.

While it is convenient for us to apportion our difficulties to others, in most cases the more relevant meaning of the twelfth is as the house of self-undoing – of the daft things we do to mess up our own lives, or, in traditional parlance, of Sin. It is a common error among students of horary to attempt to drag the eleventh house – 'hopes and wishes' – into any question, usually from a well-meaning desire to provide a positive answer. If I can't find a happy outcome from the houses concerned, maybe I can treat it as 'will my wish come true?' It is perhaps not being unduly cynical to suggest that such answers might more reasonably be found from the twelfth, as so many horary questions boil down to 'How can I make my life as difficult as possible?', the seductions of 'the World, the Flesh and the Devil' being endlessly alluring.

That they are so is shown by the planet associated with the twelfth in the Chaldean order that links, by way of the planets, the *mundane houses* to the *celestial houses* (or signs). Following this pattern round the chart from Saturn in the first house, we find Venus governing the twelfth, providing the glamour that makes temptation so tempting. Without this association, our rational minds would have no difficulty in showing us just how unpleasant our favourite means of self-undoing really are. It is, then, to the twelfth that we would look for such things as electing a time to stop smoking, or to analyse why we persist in being drawn back into some behaviour that in our more lucid moments we see as harmful.

As we become trapped in our own habits of body or of mind, shown by the twelfth, so the more material means of entrapment are shown here: prisons, and other places of confinement. In horary charts cast on cases in criminal court, the main significator entering the twelfth house is a strong indication that this person will (and the phrase brings us back to the root meaning of the twelfth, as shown by Morinus) serve time.

The twelfth is often associated with monasteries; but we need to be cautious here. In the past, packing some troublesome hussy off to a nunnery may have provided a comparatively benign form of imprisonment, and as such this act could rightly be taken as a twelfth-house matter. If the monastery is seen in its true function as a house of prayer it belongs, however, to the ninth, the house that covers spiritual matters. It is just possible that we might look to the twelfth if there were a choice between entering an open or a closed order; but we would hope that such a candidate would find better means of resolving such an issue than consulting an astrologer.

The texts do tell us that the twelfth is associated with 'monkery' in matters concerning the native's prospects for marrying or having children. This has more to do with the tendency to celibacy than with any religious vocation, however.

A final nuance on this theme is given by the early writer Paul of Alexandria. He tells us that the twelfth is the house of childbed. The old term 'confinement' makes the connection plain, and, following the order of the houses around the chart, from childbed we arrive immediately at the Ascendant, the point of birth.

That the twelfth cannot be seen from the first house – in technical language, the two houses do not *behold* each other – allied with its general indications makes it also the house of secrets. As might be expected, this meaning of the house is particularly evident in horaries on relationship matters, most of all when there is an affair. Unless there is a good reason for it being there – the partner is in prison, perhaps – finding his significator in this house is a strong sign that something untoward is going on.

After all this 'sorrow, tribulation and affliction' it is something of a relief to find that there is at least one meaning of this house without negative implications. The twelfth is the house of 'great Cattle, as Horses, Oxen, Elephants, etc'; of animals that are generically larger than a goat, or, as Abraham ben Ezra has it, which 'serve as a mount for man'.

I eagerly await my first horary on the subject of elephants, but questions on horses do come up from time to time, as do elections for the best moment to buy a horse. Haly advises that we should judge the winner of a horse race from the twelfth. This, however, is only in the most particular circumstances. If it is my horse that is running, I will look to the ruler of the twelfth. If I am using astrology to supplement my study of form, looking to the twelfth will do nothing but empty my pockets.

Even in what might seem to be a twelfth house question, such as 'Will I make money on the horses?' we should not look to the twelfth unless we are hoping to make a profit from owning a horse. As always, we must be careful to understand what is really being asked: the question here is 'Will I make money from gambling?' This is not – except when seen as a means of self-undoing – a twelfth-house matter.

As a general rule, finding, for instance, the significator of vocation in the twelfth is unfortunate. In the specific context of 'Will I succeed as a jockey, or race-horse owner, or lion tamer?' this could be just what is required to give a positive answer.

Finding a significator in the twelfth is usually unfortunate, as out of all the houses it is when here that planets find it most difficult to act, to manifest their potential. Placement in the twelfth is a strong accidental debility. This does not necessarily mean that all is lost, as although this is a debilitating position there are various ways by which a planet may still exert itself even here.

Chief among these are mutual reception and aspect. If a planet in the twelfth is in mutual reception with another that is more favourably placed, it can work through that other planet. We have all seen the movies: Mr. Big is locked up in prison and so cannot commit the crime himself, but through his associates (mutual reception) on the outside he can still arrange for the job to be done. A close aspect can work almost as well, though some degree of reception between the two planets makes the contact significantly more effective.

A good example of the way in which a twelfth house planet can still act is seen in the nativity of Tony Blair. The ruler of his Ascendant, Mercury, is shut away in the twelfth. This would usually be a negative testimony for any achievement in the world. Becoming Prime Minster is an achievement of some moment, so we would expect to find a way out for this trapped planet.

Way out there is, through a powerful mutual reception with Mars, placed exactly on the Ascendant – an excellent place from which to act. What we see here is that Mr. Blair's own nature – the Ascendant ruler – is kept well hidden, while he works through another planet that is in the place of the personal nature (the Ascendant). So we find a carefully cultivated public persona, while the real personality is as carefully concealed.

There is an apparent contradiction between the accidental weakness of a planet in the twelfth and the accidental strength that Saturn gains when

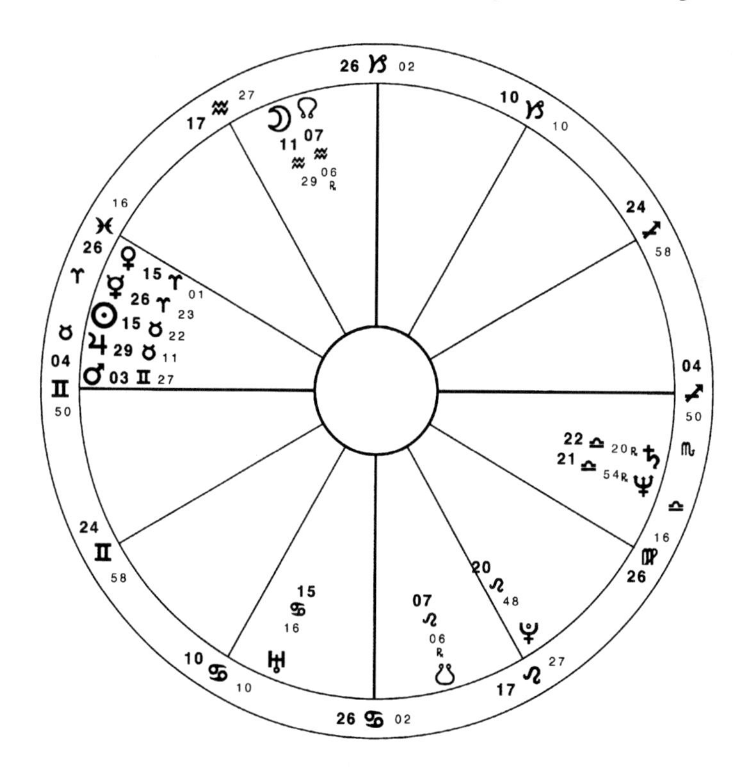

Chart 15. Tony Blair. May 6th 1953. 6.10 am BST. Edinburgh.

placed there, by being in the house of its joy. How can being in this house make a planet at once weaker and stronger? This seems to manifest as a sort of backs (Saturn) to the wall (Saturn) attitude: the gaining of strength through adversity. The army may not wish to be pushed back to the last citadel, but once it is there it is in a position from which it cannot easily be dislodged.

Lilly states that Saturn joys here 'for naturally Saturn is author of mischief', and as ever the planetary joy shows us how the nature of the house tends to manifest in our lives, in this case as the confinement and restriction of either prison bars or the yet darker cell of our own bad habits and foolishness. But the planetary joys are also our guidebook on how to deal with the matters of the relevant house. The bodily part associated with the twelfth house is the feet, with their proverbial readiness to hasten to do evil. That Saturn joys here shows that we can either let our feet lead us where they will, into toils and limitation, or we can use the virtues of Saturn, discipline and restraint, to guide them where we know they do well to tread.

NEPTUNIA REPLIES...

—a final word from our sensitive seer

Dear Neptunia, I am so confused; I know only you can help me. My boyfriend tells me that we live on an insignificant planet circling a mediocre star, and that space is so huge that we cannot possibly be important enough to warrant the receipt of any planetary influence. What should I do?

Yours despairingly, Tracey

Dear Tracey, I assume that you have tried the obvious, and thrown a bucket of water over him. This can work in the early stages of this condition, although you will soon find that his views are expressed with so much choler that any water thrown turns merely into steam.

It sounds to me as if he has been getting into bad company, talking to those who see no further than their abacus and base their philosophies on these cold beads alone. Yes, Tracey, the universe is very big; and even Alfonso, whom as I write I watch through my study window, emerging from beneath the bonnet of my Ferrari where he has been beefing the acceleration, is for all his impressive musculature, very small. But dimension alone means nothing. What is of significance is the infinite fecundity that fills this cold space. Our universe is not the vast emptiness of Saturnian dimension, but the endlessly rich intricacy of Jupiterian abundance. While looking outward, we may see it is very big; if we gaze instead downwards we cannot but marvel at the sheer much-ness with which this space is filled. As I see the sunlight sparkling through the tiny beads of sweat on Alfonso's biceps, that this *is* is sufficient proof of our centrality in a meaningful cosmos.

So I suggest you use something solid to block your boyfriend's ears until he has learned to use them with more discrimination, and encourage him instead to use his eyes, turning them downwards to this world as often as upwards to the stars. For thence alone lies comprehension of the web that binds the 'as above' and 'so below' as one.

Your caring, Neptunia

Reading and Reference

This list is limited to those works of particular relevance to the articles contained here.

ASTROLOGICAL TEXTS.

The three key works:

William Lilly, *Christian Astrology*, (1645). The best practical introduction to the subject. It is available as *Christian Astrology Books I and II*, Ascella, Nottingham, 1999, which covers the general introduction and horary section, and *Book III,* Ascella, London, 2001, which deals with natal astrology. Book III assumes a working knowledge of the techniques taught in Books I and II.

Abu 'Ali Al-Khayyat, *The Judgments of Nativities,* (9th Century) trans. James H. Holden, American Federation of Astrologers, Tempe, 1988. The clearest, soundest and most concise of the readily available texts on natal astrology.

Titus Burckhardt, *Mystical Astrology According to Ibn 'Arabi,* Beshara, Abingdon, 1977. As indispensable as Lilly for practice is this for theory. Brief, but broad in its scope, it explains the cosmological basis for astrology. A new translation is now available from Fons Vitae (www.fonsvitae.com)

Others worthy of study:

Abu'l-Rayhan Muhammad Ibn Ahmad Al-Biruni, *The Book of Instruction in the Elements of the Art of Astrology,* (1029); trans. R. Ramsey Wright, Luzac, London, 1934. Reprinted Ascella, Nottingham, n.d. A compendium of astrological knowledge, full of valuable information.

Al-Kindi, *On the Stellar Rays,* trans. Zoller, Golden Hind, Berkeley Springs, 1993. A valuable discussion of areas of the background philosophy.

Abraham ben Ezra, *The Beginning of Wisdom,* Ascella, Nottingham, n.d., and *The Book of Reasons,* trans. Epstein, Golden Hind, Berkeley Springs, 1994. Ben Ezra's exposition of the basics is useful.

Henry Coley, *Key to the Whole Art of Astrology,* (1676); reprinted Ascella, Nottingham, n.d. Much of Coley's book reproduces what is better expressed in the work of his mentor, William Lilly; but in addition to this, there is a useful section on electional astrology, some valuable chapters on nativities, and the aphorisms (*Centiloquum*) of both Ptolemy and Hermes Trismegistus. These are key texts in astrology's history.

Morinus, *The Cabal of the Twelve Houses,* Ascella, Nottingham, n.d. The author's theory of the meanings of the mundane houses.

Claudius Ptolemy, *Tetrabiblos,* trans. F.E. Robbins, Heinemann, London, 1940. Although far from comprehensive, this is the most influential book in the history of astrology and must be read by anyone who claims an interest in the subject.

Vivian E. Robson, *The Fixed Stars and Constellations in Astrology,* (1923), Ascella, Nottingham, n.d. The standard text on fixed stars.

Richard Saunders, *Astrological Judgement & Practice of Physick,* (1677), reprinted Ascella, London 2001, is the most important of the medical texts. Lilly's *Christian Astrology I & II* has much of value on this subject, while Nicholas Culpeper's *Astrological Judgment of Diseases,* (1655), Ascella, Nottingham, n.d., is as entertaining as it is informative.

BACKGROUND

History:

For the world in which Lilly moved, the best I have found is the work of Christopher Hill, which combines illumination with readability. His *Milton and the English Revolution,* Faber, London, 1977, and *The Experience of Defeat,* Faber, London, 1984, add greatly to our understanding of Lilly. His *The World Turned Upside Down,* Temple Smith, 1972, also merits study.

Thomas S. Kuhn, *The Copernican Revolution,* Harvard University Press, 1957, is the clearest account of the Ptolemaic model of the Solar System, and a useful corrective to common misconceptions about the rise of the Copernican model. Compulsory reading for anyone interested in traditional astrology.

Philosophy:

The Myth of Er at the end of Plato's *Republic* is a remarkable astrological statement. The texts assume a familiarity with Macrobius' *Commentary on the Dream of Scipio*; the Columbia University Press edition also contains Cicero's *Dream of Scipio.* Robert Grossteste's *Hexaëmeron,* available as *On the Six Days of Creation,* Oxford University Press, 1996, gives detailed explanations of many of the fundamental concepts. Ficino's *Commentary on Plato's Symposium*, Spring Publications, Woodstock, 1985 is endlessly rewarding, while the less familiar works of Dante – the *Convito* and *On Monarchy* – together, of course, with the *Comedy* illuminate many areas of traditional thought.

Of modern works, particularly worth of note are Titus Burckhardt's *Alchemy,* Watkins, London, 1967, which in its profound treatment of its subject contains much of direct astrological relevance, and Ghazi bin Muhammed's *The Sacred Origin of Sports and Culture,* Fons Vitae, Louisville, 1998, highly recommended to anyone with an interest in traditional ideas in the modern world.

ALSO BY JOHN FRAWLEY

and published by Apprentice Books

THE REAL ASTROLOGY

Winner of the Spica Award for International Book of the Year, *The Real Astrology* provides a searching – and often hilarious – critique of modern astrology and a detailed introduction to the traditional craft. It contains a clear exposition of the cosmological background and a step-by-step guide to method, accessible to those with no prior knowledge of the subject, yet sufficiently thorough to serve as a *vade mecum* for the student or practitioner.

Philosophically rich – genuinely funny – written by a master of the subject and informed with invaluable practical advice. – *The Mountain Astrologer*

Wit, philosophy and a thoroughly remarkable depth of scholarship. I will be ever thankful to John Frawley for this gem of a book. – *AFI Journal*

Required reading for all astrologers - *Prediction*

Lightning Source UK Ltd.
Milton Keynes UK
01 March 2011

168473UK00001B/43/P

9 780953 97741